AF350606

TEXT BOOK OF

ORGANIC CHEMISTRY

(Paper VIII : DSC - D4)
B.Sc. Part II : Semester IV

As per New Revised Choice Based Credit System (CBCS) Syllabus of Shivaji University, Kolhapur, w.e.f. June 2019

Dr. PRAVINA B. PISTE
M.Sc., Ph.D.
Associate Professor at Y.C. Institute
of Science, Satara.
PG Recognized Teacher, Research Guide.

Dr. ARJUN S. KUMBHAR
M.Sc., Ph.D.
Associate Professor,
P.G. Recognized Teacher, Research Guide,
P.D.V.P. College, Tasgaon. Dist. - Sangli

Dr. DNYANDEV N. ZAMBARE
M.Sc., Ph.D.
Professor,
P.G. Recognized Teacher, Research Guide,
Department of Chemistry,
Kisanveer Mahavidyalaya, Wai.

Dr. AVINASH M. NALAWADE
M.Sc., M.Phil, Ph.D.
Assistant Professor,
P.G. Recognized Teacher, Research Guide,
Lal Bahadur Shastri College, Satara

Dr. C. P. MANE
M.Sc., Ph.D.
Head, P.G. Recognized Teacher,
Research Guide, Department of Chemistry,
Lal Bahadur Shastri College, Satara.

N4829

B.Sc. - II : ORGANIC CHEMISTRY (P-VIII) (Sem. - IV)

First Edition : December 2019

ISBN 978-93-89686-68-5

© : Authors

Published By : **Polyplate**
NIRALI PRAKASHAN
Abhyudaya Pragati, 1312, Shivaji Nagar,
Off J.M. Road, PUNE – 411005
Tel - (020) 25512336/37/39, Fax - (020) 25511379
Email : niralipune@pragationline.com

➤ **DISTRIBUTION CENTRES**

PUNE

Nirali Prakashan : 119, Budhwar Peth, Jogeshwari Mandir Lane,
Pune 411002, Maharashtra. Tel : (020) 2445 2044,
Email : bookorder@pragationline.com,
niralilocal@pragationline.com

Nirali Prakashan : S. No. 28/27, Dhyari, Near Pari Company, Pune 411041
Tel : (020) 24690204 Fax : (020) 24690316
Email : dhyari@pragationline.com,
bookorder@pragationline.com

MUMBAI

Nirali Prakashan : 385, S.V.P. Road, Rasdhara Co-op. Hsg. Society Ltd.,
Girgaum, Mumbai 400004, Maharashtra
Tel : (022) 2385 6339 / 2386 9976, Fax : (022) 2386 9976
Email : niralimumbai@pragationline.com

➤ **DISTRIBUTION BRANCHES**

JALGAON

Nirali Prakashan : 34, V. V. Golani Market, Navi Peth, Jalgaon 425001,
Maharashtra, Tel : (0257) 222 0395, Mob : 94234 91860

KOLHAPUR

Nirali Prakashan : New Mahadvar Road, Kedar Plaza, 1st Floor Opp. IDBI Bank
Kolhapur 416 012, Maharashtra. Mob : 9850046155

NAGPUR

Nirali Prakashan : Above Maratha Mandir, Shop No. 3, First Floor,
Rani Jhanshi Square, Sitabuldi, Nagpur 440012, Maharashtra
Tel : (0712) 254 7129;
Email : niralinagpur@pragationline.com

DELHI

Nirali Prakashan : 4593/15, Basement, Agarwal Lane, Ansari Road, Daryaganj
Near Times of India Building, New Delhi 110002
Mob : 08505972553

BENGALURU

Nirali Prakashan : Maitri Ground Floor, Jaya Apartments, No. 99, 6th Cross,
6th Main, Malleswaram, Bangaluru 560 003, Karnataka
Mob : +91 9449043034
Email: niralibangalore@pragationline.com

Other Branches : Hyderabad, Chennai

niralipune@pragationline.com | www.pragationline.com

Also find us on   www.facebook.com/niralibooks

PREFACE

This book is basically intended for B.Sc.-II students for Semester - IV Course of Shivaji University, Kolhapur. This book is written according to New Revised Choice Based Credit System (CBCS) syllabus being implemented from June 2019 prescribed by Shivaji University, Kolhapur.

It is our great pleasure to present this book to the students and respected teachers in proper time. The subject matter is presented in simple and lucid language. This book covers all the chapters mentioned in the syllabus. The material is presented in a comprehensive way and the sequence of articles in each chapter helps the students to understand the subject with ease.

Different diagrams and illustrative description is provided to enhance the learning and understanding of the matter by students as well as to enable the teachers to explain the difficult concepts properly. The solved numerical examples, long answer type questions, short answer type questions including multiple type questions are given at the end of each chapter.

We are thankful to Nirali Prakashan, Pune for making us a part of their team of Authors. We thank Mr. **Dineshbhai Furia** and **Mr. Jignesh Furia** for publishing this book.

We also thank Mr. Girish Redkar (Head Marketing Dept.) for his co-operation in publishing this book.

Last but not the least we are very much indebted to Mr. Virdhaval Shinde (Marketing Executive, Kolhapur District) and Mr. Ashok Nanavare (Marketing Executive, Sangli District) for their nice co-operation. We are very much thankful to Mr. Kiran Velankar (Proof Reading), Mrs. Anjali Muley (Graphic Design) and Mr. Malik Shaikh for a neat and error free D.T.P. of this book.

We hole that this book will be found useful by students and teachers. We will appreciate any suggestions for the improvement of the book.

– **Authors**

Unit 1 : Carboxylic Acids and their Derivatives (8 L)

1.1 Monocarboxylic acid: Introduction, Methods of Formation from Alcohols, Aldehydes, Ketones, Nitriles and Alkyl benzenes. Chemical Reaction: Hell-Volhard-Zelinsky (HVZ) reaction.

1.2 Formation of Halo Acids: Mono, Di, Tri- chloro acetic acid. Substitution reactions of Monochloro acetic acid by Nucleophile OH^-, I^-, CN^- and NH_3

1.3 Hydroxy acids: Malic and Citric acid. Methods of formation of Malic acid from maleic acid, from Alpha bromo succinic acid and moist Ag_2O. Chemical Reactions: Reactions of Malic acid-Action of heat, oxidation by $KMnO_4$ and reduction reaction with HI. Uses of Malic acid. Method of formation of Citric acid from glycerol. Chemical Reactions: Reaction of citric acid: Acetylation by acetic anhydride, reduction by HI, action of heat. Uses of citric acid.

1.4 Unsaturated acid: Cinnamic acid: Methods of formation from benzaldehyde using diethyl malonate and by using acetic anhydride and sodium acetate. Chemical Reactions-Bromination, Oxidation. Uses of cinnamic acid. Acrylic acid: Method of formation from acrolein and by dehydration of beta hydroxy propionic acid. Chemical Reactions: Addition of water, Reduction by Na/C_2H_5OH. Uses of acrylic acid.

1.5 Dicarboxylic acid: Succinic and phthalic acid. Method of formation of succinic acid from ethylene dibromide, maleic acid. Chemical Reactions: Action of heat, Action of $NaHCO_3$, C_2H_5OH in the presence of acid. Uses of succinic acid. Phthalic acid: Method of formation from o-xylene and Naphthalene. Chemical Reactions: Action of heat, reaction with sodalime, ammonia. Uses of phthalic acid.

1.6 Carboxylic acid derivatives: Introduction Acid halide derivative: Acetyl chloride: formation from acid, by action with PCl_3 and $SOCl_2$, reaction with water, alcohol (Mechanism of esterification is expected) and ammonia. Uses of acetyl choride. Acid anhydride derivative: Method of formation of acetic anhydride by dehydration of acetic acid, reactions with water, alcohol and ammonia. Uses of acetic anhydride.

Unit 2 : Amines and Diazonium Salts (8 L)

2.1 Introduction, Classification, Nomenclature, Structure.

2.2 Methods of preparation: (a) From alkyl halide by ammonolysis, (b) By reduction of nitriles or cyanides, (c) From unsubstituted amides (Hoffmann degradation), (d) By Gabriel synthesis (from phthalimide).

2.3　Reactions: Carbylamine reaction, Schotten-Baumann reaction, Electrophilic substitution (Aniline), Nitration, Bromination, Sulphonation.

2.4　Diazonium salt: Introduction, Preparation of Benzene diazonium chloride.

2.5　Reactions: Replacement by Halogen (Sandmeyer), Replacement by Iodine, Replacement by –OH, C and N. Coupling reactions: Synthesis of Methyl Orange and Congo Red. Reduction of BDC.

Unit 3 : Carbohydrates　　　　　　　　　　　　　　(8 L)

Classification of carbohydrates, Reducing and non-reducing sugars, General properties of glucose and fructose, their open chain structure. Epimers, mutarotation and anomers.

Determination of configuration of Glucose (Fischer proof). Ring structure of glucose. Determination of size of the ring of Glucose by methylation method. Haworth projections. Cyclic structure of fructose. Linkage between monosachharides, structure of disaccharides (sucrose, maltose, lactose) and polysaccharides (starch and cellulose), excluding their structure elucidation.

Unit 4 : Carbonyl Compounds - Aldehydes and Ketone　(6 L)

4.1　Introduction, Nomenclature, Structure.

4.2　Reactivity of Carbonyl group, Mechanism of nucleophilic addition to carbonyl group.

4.3　Reactions: Mechanism and application of Aldol condensation, Perkin reaction, Cannizzaro's reaction, Knoevenagel condensation, Reformatsky reaction.

Unit 5 : Stereochemistry　　　　　　　　　　　　　(8 L)

5.1　Conformational isomerism – Introduction.

5.2　Representation of conformations of ethane by using Saw-Horse, Fischer (dotted line wedge) and Newman's projection formulae.

5.3　Conformations and conformational analysis of ethane and n-butane by Newman's projection formula with the help of energy profile diagrams.

5.4　Cycloalkanes relative stability - Baeyer's strain theory, Theory of strainless rings.

5.5　Conformations and stability of cyclohexane and monosubstituted cyclohexanes: Cyclohexanol, bromocyclo-hexane and methyl cyclohexane.

5.6　Locking of conformation in t-butyl cyclohexane.

✱✱✱

CONTENTS

***

B.Sc. II Revised Syllabus 2018-19

Semester III and IV Nature of Question Paper

Total Marks 50

Q. 1 (a) Answer the following in one sentence: (5)

i)

ii)

iii)

iv)

v)

(b) Choose the correct alternative and rewrite the sentence again: (5)

I)

II)

III)

IV)

V)

Q. 2 Attempt any TWO of the following (Out of FOUR): (20)

a)

b)

c)

d)

Q. 3 Answer any FOUR of the following (Out of SIX): (20)

a)

b)

c)

d)

e)

f)

***

Chapter **1** ...

Carboxylic Acids and Their Derivatives

Contents ...

Introduction

It is a group of an organic compound containing a carboxylic group (–COOH). Carboxylic acid contains a carbonyl group to which the hydroxyl group is attached. The general formula of the group is R–COOH. In the formula, R denotes the rest of the group attached to the functional group. This carboxylic group may be attached to hydrogen (HCOOH), an alkyl group (RCOOH), or an aryl group (ArCOOH). The structure of carboxylic acid is

$$R-\overset{\displaystyle O \atop \displaystyle \|}{C}-OH$$

HCOOH	CH_3COOH	$CH_3(CH_2)_{10}COOH$
Formic acid	Acetic acid	Lauric acid
(Methanoic acid)	(Ethanoic acid)	(Dodecanoic acid)

Benzoic acid

Phenyl acetic acid

Cyclohexane carboxylic acid

The classification of carboxylic acids is based on the number of –COOH groups present in the molecule as mono-, di-, tri-, or poly carboxylic acid. If carboxylic acid contain halogen atom then it is called as haloacid and if it contain –OH group then it is called as hydroxy acid. When it contain double bond then it is called as unsaturated acid.

1.1 Monocarboxylic Acid

The molecule which contain only one –COOH group is called as monocarboxylic acid.

Lower straight chain aliphatic carboxylic acids as well as those of even carbon upto C_{18} are commercially available. e.g. acetic acid is produced by methanol carbonylation with carbon monoxide. Whereas long chain carboxylic acids are obtained by the hydrolysis of triglycerides obtained from plant or animal oils. Vinegar, a dilute solution of acetic acid is biologically produced from the fermentation of ethanol.

The aliphatic carboxylic acids have been known from the ancient time and named according to their sources rather than to their chemical structure. e.g. Formic acid adds the sting to the bite of an ant (Latin : Formica – ant), butyric acid gives rancid butter its typical smell (Latin : butyrum – butter); and caproic, caprylic and capric acids all are found in goat fat (Latin : Caper – goat).

Some of the examples of carboxylic acids have been given below in Table 1.1.

Table 1.1

Name of the compound	Common name	IUPAC name
HCOOH	Formic acid	Methanoic acid
CH_3COOH	Acetic acid	Ethanoic acid
$CH_3.CH_2.COOH$	Propionic acid	Propanoic acid
$CH_3.CH_2.CH_2.COOH$	Butyric acid	Butanoic acid

1.1.1 Methods of Formation

There are different methods used in preparation of monocarboxylic acids. Some of them are,

1. From Alcohols: Primary alcohols can undergo oxidation reaction to form corresponding carboxylic acids with the help of oxidizing agents such as potassium permanganate ($KMnO_4$ for neutral or acidic or alkaline media), chromium trioxide (CrO_3 /H_2SO_4 Jones reagent), and potassium dichromate ($K_2Cr_2O_7$ acidic media).

$$CH_3{-}CH_2{-}OH \xrightarrow[\substack{^-OH \\ Heat \\ 2.\ H_2O}]{1.\ KMnO_4} CH_3{-}\overset{\overset{\textstyle O}{\|}}{C}{-}OH$$

Ethanol   Ethanoic acid

2. From Aldehydes: Aldehydes with mild oxidizing agents such as Tollen's reagents [$Ag(NH_3)_2^+OH^-$] and manganese dioxide (MnO_2) give monocarboxylic acid.

$$CH_3CHO \xrightarrow{\text{Tollen's reagent}} CH_3COOH$$

Acetaldehyde   Acetic acid

3. Ketones: Methyl ketones can be converted to carboxylic acids via. the haloform reaction.

$$Ar-\overset{O}{\underset{\|}{C}}-CH_3 \xrightarrow[\quad 2.\ H_3O^+ \quad]{1.\ X_2/NaOH} Ar-\overset{O}{\underset{\|}{C}}-OH + CHX_3$$

4. Hydrolysis of Nitriles (Cyanides): Aliphatic nitriles are prepared by treatment of alkyl halides with sodium cyanide in a solvent that will dissolve both reactants. In dimethyl sulfoxide (DMSO), reaction occurs rapidly and exothermically at room temperature. The resulting nitrile is then hydrolysed to the acid by boiling with aqueous alkali or acid.

$$R-C\equiv N \xrightarrow[H^+ / HOH]{Acid\ hydrolysis} R \cdot COOH + NH_2$$

Example:

$$Br-CH_2-CH_2-CH_2-Br \xrightarrow{NaCN} NC-CH_2-CH_2-CH_2-CN$$

$$\text{Pentane dinitrile}$$

$$\xrightarrow{H_3O^+} HO_2C-CH_2-CH_2-CH_2-CO_2H$$

$$\text{Glutaric acid}$$

This synthetic method is generally limited to the use of primary alkyl halides. Aryl halides (except for those with ortho and para nitro groups) do not react with sodium cyanide. Just like Grignard synthesis, nitrile synthesis also increases the length of carbon chain.

5. Oxidation of Alkyl benzenes: Strong oxidation of alkyl benzenes also result in the formation of carboxylic acids.

$$Ar-R \xrightarrow{KMnO_4} Ar-COOH$$

Example: $NO_2-C_6H_5-CH_3 \xrightarrow{KMnO_4} NO_2-C_6H_5-COOH$

1.1.2 Chemical Reaction

1. Hell-Volhard-Zelinski reaction: Aliphatic carboxylic acids on reaction with bromine in the presence of phosphorus produce halo acids. This reaction is known as Hell-Volhard-Zelinski (HVZ) reaction.

$$CH_3CH_2COOH \xrightarrow{Br_2/Red\ P} CH_3-\underset{\underset{Br}{|}}{CH}-\overset{O}{\underset{\|}{C}}-OH$$

Propanoic acid　　　　　　　　2-Bromopropanoic acid

1.2 Halo Acids

If acid contain halogen atom then it is called as Halo acid. Depending upon the position of halogen atom on α, β, γ carbon atom with respect to carboxylic acid, they are classified as α-halo acid, β-halo acid, γ-halo acid respectively.

1.2.1 Methods of Formation of Halo Acids

Synthesis of mono-, di- and trichloro acetic acid have been carried out by different methods.

(A) Methods of formation of Monochloro acetic acid:

It can be prepared with the help of following methods:

1.   Hell-Volhard-Zelinski reaction: The reaction of aliphatic carboxylic acids with bromine/chlorine in the presence of red phosphorus in direct sunlight at 373 K produces α halo acids. This reaction is **Hell-Volhard-Zelinski reaction**.

$$CH_3COOH + Cl_2 \xrightarrow[\text{Direct sunlight}]{\text{Red P / 373 K}} Cl \cdot CH_2 - COOH + HCl$$

Acetic acid   Monochloro acetic acid

2.   From Ethylene chlorohydrine: This method is used in laboratory or in industry. The oxidation of ethylene chlorohydrine with nitric acid gives monochloro acetic acid.

$$OH - CH_2 - CH_2 - Cl \xrightarrow[\text{HNO}_3]{[O]} Cl \cdot CH_2 - COOH$$

Ethylene chlorohydrine   Monochloro acetic acid

3.   From Trichloro ethylene: When trichloro ethylene agitating with 90% sulphuric acid gives monochloro acetic acid.

$$Cl_2C = CHCl \xrightarrow{\text{90\% H}_2\text{SO}_4} Cl \cdot CH_2 - COOH$$

Trichloro ethylene   Monochloro acetic acid

(B) Dichloro acetic acid:

It is a colourless liquid having boiling point 193°C. It is soluble in water and ethanol and having 1.25 pK_a value.

Methods of Formation:

It has been prepared by following methods:

1.   Hell-Vohlard-Zelinsky method: When acetic acid is treated with excess amount of chlorine in direct sunlight in the presence of red phosphorus at 373 K, it gives dichloro acetic acid.

$$CH_3COOH + 2Cl_2 \xrightarrow[\text{Direct sunlight}]{\text{Red P / 373 K}} Cl_2 \cdot CH - COOH + 2HCl$$

Acetic acid   Dichloro acetic acid

2.   From Chloral hydrate: When chloral hydrate is treated with calcium carbonate in the presence of sodium cyanide, it gives Ca-salt of acid which on further acidification with H_2SO_4 gives dichloro acetic acid.

$$2CCl_3\text{-}CH(OH)_2 + 2CaCO_3 \xrightarrow{\text{NaCN}} (Cl_2CHCOO)_2Ca + CaCl_2 + H_2O + 2CO_2$$

Chloral hydrate

$$H_2SO_4 \downarrow H^+$$

$$2Cl_2CH.COOH + CaSO_4$$

Dichloro acetic acid

(C) Trichloro acetic acid:

It is a colourless deliquescent solid having melting point 57.5°C. It is soluble in water, ethanol and ether. It is strongly corrosive to skin.

Methods of Formation:

It has been prepared by following methods:

1.   Hell-Vohlard-Zelinsky method: When acetic acid is treated with excess amount of chlorine in direct sunlight in the presence of red phosphorus at 373 K, it gives trichloro acetic acid.

$$CH_3COOH + 3Cl_2 \xrightarrow[\text{Direct sunlight}]{\text{Red P / 373 K}} Cl_3C - COOH + 2HCl$$

Acetic acid   Trichloro acetic acid

2.   When oxidation of chloral with nitric acid is carried out it gives trichloro acetic acid.

$$CCl_3CHO \xrightarrow[\text{HNO}_3]{[O]} Cl_3C - COOH$$

Chloral   Trichloro acetic acid

1.2.2 Substitution Reactions of Monochloro Acetic Acid

1. **By Nucleophile OH⁻:** When chloroacetic acid is treated with NaOH it gives hydroxyl acetic acid (Glycolic acid).

$$R-\underset{\underset{\displaystyle Cl}{|}}{CH}-COOH \xrightarrow{\text{NaOH}} R-\underset{\underset{\displaystyle OH}{|}}{CH}-COOH + NaCl$$

$$CH_3-\underset{\underset{\displaystyle Br}{|}}{CH}-COOH + AgOH \xrightarrow{\text{NaOH}} CH_3-\underset{\underset{\displaystyle OH}{|}}{CH}-COOH + AgBr$$

2-Bromopropanoic acid   Lactic acid

2. **With I⁻:** When chloroacetic acid is treated with potassium iodide in ethanol, it gives iodoacetic acid.

$$Cl \cdot CH_2-COOH + KI \xrightarrow{\text{ethanol}} ICH_2COOH$$

Monochloro acetic acid   Iodoacetic acid

3. **With CN⁻:** When monochloro acetic acid is treated with KCN in ethanol, it gives cyanoacetic acid.

$$Cl \cdot CH_2-COOH + KCN \xrightarrow{\text{ethanol}} CNCH_2COOH$$

Monochloro acetic acid   Cyanoacetic acid

4. **With NH₃:** When chloro acetic acid is treated with ammonia, it gives glycine.

$$Cl \cdot CH_2-COOH + 2NH_3 \xrightarrow{\text{ethanol}} H_2N \cdot CH_2 \cdot COOH$$

Monochloro acetic acid   Amino acetic acid
(Glycine)

1.3 Hydroxy Acids

When the acid contains hydroxyl group (–OH), it is called as hydroxy acid. The hydroxyl acids can be classified according to number of (–OH) groups present in the molecule as mono-, di-, tri-, or polyhydroxy acids depending upon the number of –one, two, three or more respectively.

Table 1.2

Structure	Common name	IUPAC name
HO-CH$_2$-COOH	Hydroxy acetic acid	Glycolic acid

H$_3$C—CH—COOH
       |
      OH
	Hydroxy propanoic acid	Lactic acid

HOOC—CH$_2$—CH—COOH
           |
          OH
	Hydroxy butanoic acid	Malic acid

H$_3$C—HC—CH—COOH 	| OH OH	Dihydroxy butanoic acid	Tartaric acid

HOOC   COOH
  \ |
  CH$_2$—C—CH$_2$
 	
	\COOH
	     OH
		3-carbonyl, 3-hydroxyl pentane 1,5-dioic acid	Citric acid

1.3.1 Malic Acid

Malic acid is an organic compound with the molecular formula C$_4$H$_6$O$_5$. It is also known as hydroxyl butanedioic acid. It is a dicarboxylic acid that is made by all living organisms, contributes to the pleasantly sour taste of fruits, and is used as a food additive. Malic acid is a colourless crystalline solid, soluble in water and alcohol but sparingly soluble in ether, it melts at 130°C. Malic acid contains one asymmetric carbon, hence it exists in two optically active forms (two stereoisomeric forms: L- and D-enantiomers) and one inactive form, though only the L-isomer exists naturally.

It is a colourless deliquescent solid. It is showing optical isomerism. When it is present in dextro and laevo form then it has melting point 373 K and when it is present in racemic form it has melting point 408 K.

$$HOOC-CH_2-CH-COOH$$
$$|$$
$$OH$$

Malic acid

Methods of Formation:

1.   **From Maleic acid:** On hydrolysis of maleic acid, it gives malic acid.

$$HOOC-CH=CH-COOH \xrightarrow[H_2SO_4]{H^+} HOOC-CH_2-\underset{\underset{OH}{|}}{CH}-COOH$$

Maleic acid   Malic acid

2.   From α-bromo succinic acid and moist Ag$_2$O: On hydrolysis of α-bromo succinic acid and moist Ag$_2$O, it gives malic acid.

$$HOOC-CH_2-\underset{\underset{Br}{|}}{CH}-COOH \xrightarrow[H_2SO_4]{H^+} HOOC-CH_2-\underset{\underset{OH}{|}}{CH}-COOH + AgBr$$

α-Bromo succinic acid   Malic acid

Chemical Reactions:

1.   Action of heat: When malic acid is heated at 423 K to 452 K, it gives maleic acid.

$$HOOC-CH_2-\underset{\underset{OH}{|}}{CH}-COOH \xrightarrow[423\ K - 452\ K]{\Delta} HOOC-CH=CH-COOH$$

Malic acid   Maleic acid

2.   Oxidation by KMnO$_4$: On oxidation of malic acid, it gives oxaloacetic acid.

$$HOOC-CH_2-\underset{\underset{OH}{|}}{CH}-COOH \xrightarrow[KMnO_4]{[O]} HOOC-CH_2-\overset{\overset{O}{\|}}{C}-COOH$$

Malic acid   Oxalo acetic acid
                                                        (Keto form)

$$HOOC-CH=\underset{\underset{OH}{|}}{C}-COOH$$

(Enol form)

3.   Reduction reaction with HI: When malic acid is treated with HI, it reduces malic acid and gives succinic acid.

$$HOOC-CH_2-\underset{\underset{OH}{|}}{CH}-COOH \xrightarrow{2HI} HOOC-CH_2-CH_2-COOH$$

Malic acid   Succinic acid

Uses of Malic Acid:
1. It is used as beverages.
2. It is used as a remedy for throat.
3. It is also used as a substitute for citric acid.
4. It is also used as medicine as purgative.
5. It is also used as food acidulent.

1.3.2 Citric Acid

Citric acid is present in lemon, orange and other fruits. In lemon juice, it is present at about 6-10%. The structure of citric acid is given below.

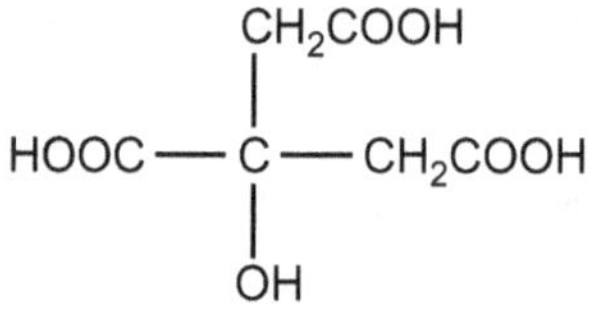

$$CH_2COOH$$
$$|$$
$$HOOC-C-CH_2COOH$$
$$|$$
$$OH$$

Citric acid

Method of Formation of Citric Acid:

1.   From glycerol by Grimaux and Adam synthesis (1880): Glycerol shows citric acid by following series of reactions. The sequential steps in this synthesis are:

1,2,3-hydroxy propane → 1,3-dichloro-2-propenol → 1,3-dichloro-2-propanone → 1,3-dichloro-2-cyano-2-hydroxypropane → 1,3-dichloromethyl, 2-hydroxypropionic acid → 1,3-dicyanomethyl-2-hydroxypropionic acid → 2-Hydroxypropane-1,2,3-tricarboxylic acid (Citric acid).

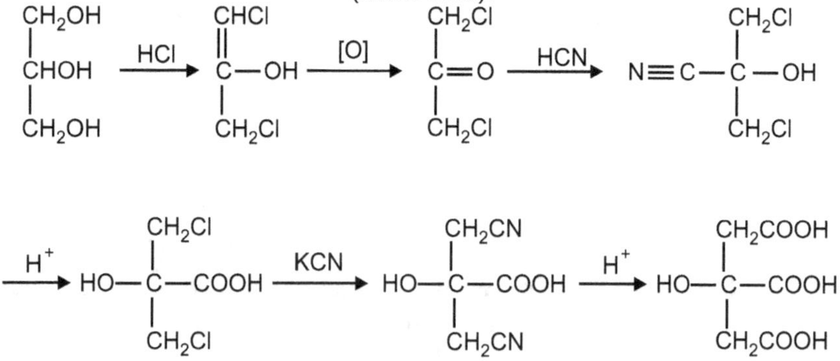

Chemical Reactions:

1.   Acetylation by acetic anhydride: Citric acid produces monoacetyl derivatives.

$$\underset{\text{Citric acid}}{\begin{array}{c}CH_2COOH\\|\\C(OH)COOH\\|\\CH_2COOH\end{array}} + \underset{\text{Acetic anhydride}}{(CH_3CO)_2O} \longrightarrow \underset{\text{Monoacetyl citric acid}}{\begin{array}{c}CH_2COOH\\|\\CH_3COO-C-COOH\\|\\CH_2COOH\end{array}}$$

2.   **Reduction by HI:** In the presence of hydrogen iodide, citric acid is reduced to tricarboxylic acid.

$$\underset{\text{Citric acid}}{\begin{array}{c}CH_2COOH\\|\\C(OH)COOH\\|\\CH_2COOH\end{array}} + 2HI \longrightarrow \underset{\substack{\text{Tricarboxylic}\\\text{acid}}}{\begin{array}{c}CH_2COOH\\|\\CHCOOH\\|\\CH_2COOH\end{array}} + H_2O + I_2$$

3.   **Action of heat:** On heating with 150°C, citric acid undergoes dehydration and gives aconitic acid.

$$\underset{\text{Citric acid}}{\begin{array}{c}CH_2COOH\\|\\C(OH)COOH\\|\\CH_2COOH\end{array}} \xrightarrow{150°C} \underset{\text{Aconitic acid}}{\begin{array}{c}CHCOOH\\||\\CCOOH\\|\\CH_2COOH\end{array}}$$

Uses of Citric Acid:

1.   As a flavour compound in the preparation of synthetic fruit drinks.
2.   As a laxative in the form of magnesium citrate.
3.   As a solvent in polymer synthesis.
4.   As an iron supplement in the form of ferric ammonium citrate.
5.   As a mordant in printing and dying.

1.4 Unsaturated Acid

If acid contain double bond then it is called as unsaturated acid. If double bond is conjugated with acid group then it is called as α, β-unsaturated acid.

Table 1.3

Structure of compound	Common name	IUPAC name
$H_2C = CH \cdot COOH$	Acryllic acid	Propenoic acid
$H_3C - CH = CH \cdot COOH$	Crotonic acid	But-2-enoic acid

Structure of compound	Common name	IUPAC name
⬡—CH=CH.COOH	Cinnamic acid	3-phenyl propenoic acid

1.4.1 Cinnamic Acid

It is a white solid compound having melting point 406 K.

⬡—CH=CH.COOH

Methods of Formation:

1. From benzaldehyde using diethyl malonate: When benzaldehyde is condensed with diethyl malonate in pyridine followed by hydrolysis and decarboxylation, it gives cinnamic acid. This reaction is known as **Knoevenagel condensation.**

$$\underset{\text{Benzaldehyde}}{\text{C}_6\text{H}_5\text{OHC}} + \underset{\text{Diethyl malonate}}{\text{H}_2\text{C}(\text{COOC}_2\text{H}_5)_2} \xrightarrow{\text{Pyridine}} \text{C}_6\text{H}_5-\text{CH}=\text{C}(\text{COOC}_2\text{H}_5)_2$$

$$\xrightarrow{\text{H}^+} \text{C}_6\text{H}_5-\text{CH}=\text{C}(\text{COOH})_2 \xrightarrow{-\text{CO}_2} \underset{\text{Cinnamic acid}}{\text{C}_6\text{H}_5-\text{CH}=\text{CH.COOH}}$$

2. By using acetic anhydride and sodium acetate: When benzaldehyde is treated with acetic anhydride in the presence of sodium acetate at 453 K, it gives cinnamic acid. This reaction is known as **Perkin reaction.**

$$\underset{\text{Benzaldehyde}}{\text{C}_6\text{H}_5\text{OHC}} + \underset{\text{Acetic anhydride}}{[(\text{CH}_3\text{CO})_2]\text{O}} \xrightarrow[453\ \text{K}]{\text{CH}_3\text{COONa}} \underset{\text{Cinnamic acid}}{\text{C}_6\text{H}_5-\text{CH}=\text{CH.COOH}}$$

Chemical Reactions:

1. Action with Br₂ (Bromination): The cinnamic acid on bromination with Br_2 gives di-bromo cinnamic acid.

$$C_6H_5-CH=CH.COOH \xrightarrow{\ Br_2\ } C_6H_5-CH-CH-COOH$$

Cinnamic acid → Dibromo cinnamic acid (with Br, Br substituents)

2. Oxidation:

(a) With KMnO₄: On oxidation with $KMnO_4$, cinnamic acid gives benzaldehyde, further vigorous oxidation gives benzoic acid.

$$C_6H_5-CH=CH.COOH \xrightarrow[\text{[O]}]{KMnO_4} \text{Benzaldehyde (OHC—)}$$

Cinnamic acid → Benzaldehyde

Benzaldehyde $\xrightarrow[\text{oxidation}]{\text{[O] Vigorous}}$ Benzoic acid (HOOC—)

(b) With chromic acid: On oxidation with chromic acid, it gives a mixture of benzaldehyde and benzoic acid.

$$C_6H_5-CH=CH.COOH \xrightarrow[\text{[O]}]{CrO_3} \text{Benzaldehyde (OHC—)} + \text{Benzoic acid (HOOC—)}$$

Cinnamic acid → Benzaldehyde + Benzoic acid

Uses of Cinnamic Acid:

1. It is used as a preservative.
2. It is used in the preparation of unsaturated esters and substituted unsaturated acids.
3. It is used in pharmaceutical industry and perfumery industry.

1.4.2 Acrylic Acid

It is a colourless liquid having boiling point 415 K and has pungent smell.

$$H_2C = CH \cdot COOH$$
Acrylic acid

Methods of Formation:

It has been prepared by following methods.

1.   From Acrolein: Acrolein on oxidation with ammonical solution of silver nitrate gives acrylic acid.

$$H_2C = CH \cdot CHO \xrightarrow[\text{Amm. } AgNO_3]{[O]} H_2C = CH \cdot COOH$$

Acrolein   Acrylic acid

2.   By dehydration of β-hydroxy propionic acid: β-hydroxy propanoic acid when heated with NaOH or $ZnCl_2$, it loses water molecule and gives acrylic acid.

$$OH \cdot CH_2 \cdot CH_2 \cdot COOH \xrightarrow[\Delta]{ZnCl_2 \text{ / } NaOH} H_2C = CH \cdot COOH + H_2O$$

β-hydroxy propanoic acid   Acrylic acid

Chemical Reactions:

It undergoes following reactions.

1.   Addition of water: Acryllic acid on acid hydrolysis in the presence of $HgSO_4$ gives β-hydroxy propanoic acid.

$$H_2C = CH \cdot COOH + H_2O \xrightarrow[H_2SO_4]{HgSO_4} OH \cdot CH_2 \cdot CH_2 \cdot COOH$$

Acrylic acid   β-hydroxy propanoic acid

2.   Reduction by Na/C_2H_5OH: Acrylic acid on reduction with sodium in ethanol gives propanoic acid.

$$H_2C = CH \cdot COOH \xrightarrow[C_2H_5OH]{Na} CH_3 \cdot CH_2 \cdot COOH$$

Acrylic acid   Propanoic acid

Uses of Acrylic Acid:

1.   It is used in the preparation of ester.
2.   It is used as a modifying agent for thermosetting.

1.5 Dicarboxylic Acid

It contains two carboxyl groups, one at each end of a saturated hydrocarbon chain. Its general formula is $HOOC - (CH_2)_n - COOH$, where n = 0, 1, 2, 3, etc. Dicarboxylic acids are named as alkane dioic acids in the IUPAC system. Most simple dicarboxylic acids have common names.

IUPAC Nomenclature of Carboxylic acids: If there are two –COOH groups present in an acid, then the acid is called dicarboxylic acid. To construct the IUPAC name of these compounds, add the suffix *–dioic* acid to the name of the parent alkane containing both carboxylic groups.

Table 1.4

Formula	Common name	IUPAC name
HOOC – COOH	Oxalic acid	Ethane dioic acid
HOOC – CH_2 – COOH	Malonic acid	Propane dioic acid
HOOC–$(CH_2)_2$–COOH	Succinic acid	Butane dioic acid
HOOC – $(CH_2)_4$ – COOH	Adipic acid	Hexane dioic acid

1.5.1 Succinic Acid

Butanedioic acid is HOOC–CH_2–CH_2–COOH

Methods of Formation of Succinic Acid:

1.   **From Ethylene dibromide:** Succinic acid is prepared from ethylene dibromide by treating with sodium cyanide and subsequent hydrolysis of ethylene dicyanide.

$$\underset{\text{Ethylene dibromide}}{\overset{\displaystyle CH_2Br}{\underset{\displaystyle CH_2Br}{|}}} \xrightarrow{2NaCN} \underset{\text{Ethylene dicyanide}}{\overset{\displaystyle CH_2CN}{\underset{\displaystyle CH_2CN}{|}}} \xrightarrow[\text{HCl}]{H_2O} \underset{\text{Succinic acid}}{\overset{\displaystyle CH_2COOH}{\underset{\displaystyle CH_2COOH}{|}}}$$

2.   **From Maleic acid:** The maleic acid on reduction with catalyst like Ni gives succinic acid.

$$\underset{\text{Maleic acid}}{\overset{\displaystyle HC-COOH}{\underset{\displaystyle HC-COOH}{\|}}} \xrightarrow[\text{Reduction}]{Ni} \underset{\text{Succinic acid}}{\overset{\displaystyle H_2C-COOH}{\underset{\displaystyle H_2C-COOH}{|}}}$$

Chemical Reactions:

1.   **Action of heat:** Succinic acid on heating to 300°C loses a molecule of water to form succinic anhydride.

$$\underset{}{\overset{\displaystyle CH_2-COOH}{\underset{\displaystyle CH_2-COOH}{|}}} \xrightarrow{300°C} \underset{\text{Succinic anhydride (A cyclic compound)}}{\overset{\displaystyle CH_2-CO}{\underset{\displaystyle CH_2-CO}{|}}}\!\!\diagdown\!\!\diagup O \; + \; H_2O$$

2.   Action of NaHCO₃: On reaction with sodium bicarbonate, succinic acid gives disodium succinate.

$$H_2C\!-\!COOH \quad \xrightarrow{NaHCO_3} \quad H_2C\!-\!COONa \quad \xrightarrow{NaHCO_3} \quad H_2C\!-\!COONa$$
$$H_2C\!-\!COOH \qquad\qquad\qquad H_2C\!-\!COOH \qquad\qquad\qquad H_2C\!-\!COONa$$

Succinic acid   Sodium salt of   Disodium succinate
                       succinic acid

3.   With C₂H₅OH in the presence of acid: Succinic acid when treated with ethanol in the presence of acid gives diethyl succinate.

$$H_2C\!-\!COOH \quad + 2C_2H_5OH \quad \xrightarrow[\text{Acid}]{H^+} \quad H_2C\!-\!COOC_2H_5$$
$$H_2C\!-\!COOH \qquad\qquad\qquad\qquad\qquad H_2C\!-\!COOC_2H_5$$

Succinic acid   Diethyl succinate

Uses of Succinic Acid:

1.   It is very important laboratory reagent.
2.   It is used in the manufacture of high polymer ester.
3.   It is used in the manufacture of lacquers and dyes.

1.5.2 Phthalic Acid

The structure of phthalic acid is,

COOH

COOH

Phthalic acid

Methods of Formation:

1.   From o-xylene: Oxidation of o-xylene with KMnO₄ / V₂O₅ (at 573 K) gives phthalic acid.

CH₃   COOH
          $\xrightarrow[KMnO_4]{[O]}$
CH₃   COOH
o-Xylene   Phthalic acid

CH₃   COOH
          $\xrightarrow[\substack{V_2O_5 \\ 573\ K}]{[O]}$
CH₃   COOH
o-Xylene   Phthalic acid

2.   From Naphthalene: Oxidation of naphthalene at 573 K - 773 K gives phthalic anhydride which on treatment with alkali gives disodium phthalate, further with acid hydrolysis gives phthalic acid.

Naphthalene → (V$_2$O$_5$, [O], 573 K - 733 K) → Phthalic anhydride → (2NaOH) → (COONa, COONa) → (2HCl) → Phthalic acid (COOH, COOH)

Chemical Reactions:

It undergoes different reactions such as,

1. **Action of Heat:** Phthalic acid on heating gives phthalic anhydride.

Phthalic acid (COOH, COOH) → (Δ, Heat) → Phthalic anhydride (CO, CO, O)

2. **Reaction with Sodalime:** Phthalic acid when treated with soda lime gives benzoic acid but in excess amount it gives benzene.

Phthalic acid (COOH, COOH) → (NaOH (CaO), $-CO_2$) → Benzoic acid (COOH) → (NaOH (CaO), $-CO_2$) → Benzene

3. **With Ammonia:** Phthalic acid when treated with NH_3 gives phthalimide.

Phthalic acid (COOH, COOH) → (NH_3) → Phthalimide (CO, NH, CO)

Uses of Phthalic Acid:

1. It is used in the manufacture of dyes.
2. It is used in the preparation of benzoic acid and substituted aromatic acids.

1.6 Carboxylic Acid Derivatives

The compound which is derived from acid is called as carboxylic acid derivatives.

Table 1.5

Structure	Common name	IUPAC name
$CH_3 \cdot CO \cdot Cl$	Acetyl chloride	Acid chloride
$CH_3 \cdot CO \cdot O \cdot CO \cdot CH_3$	Acetic anhydride	Acid anhydride
$CH_3 \cdot COO \cdot C_2H_5$	Ethyl acetate	Acid ester
$CH_3 \cdot CO \cdot NH_2$	Acetamide	Acid amide

1.6.1 Acetyl Chloride

Acetyl chloride is volatile, pungent smelling liquid, this cannot form hydrogen bond. Hence its low boiling point and insolubility in water is explained. It fumes in moist air due to hydrolysis producing hydrogen chloride gas. In IUPAC system, acid chlorides are named by replacing the "e" ending of the parent alkane by "-oyl chloride". The electron withdrawing inductive effect of an acyl chloride is not stabilized by electron pair donation; the electron withdrawing inductive effect of chlorine makes it more electrophilic and more reactive towards nucleophilic acyl substitution.

Method of Formation:

It has been prepared by following method.

1.   From acid by action with PCl$_3$ and SOCl$_2$: When acetic acid is treated with phosphorus trichloride or thionyl chloride it gives acetyl chloride.

(i) $CH_3COOH + PCl_5 \longrightarrow CH_3COCl + POCl_3 + HCl$

(ii) $3CH_3COOH + PCl_3 \longrightarrow 3CH_3COCl + H_3PO_3$

Phosphorus acid

(iii) $CH_3COOH + SOCl_2 \longrightarrow CH_3COCl + SO_2 + HCl$

Chemical Properties:

It is a powerful acetylating agent, a reagent that introduces CH_3CO -group, especially with compounds containing –OH group or –NH group.

1. **Reaction with water and alcohol (Mechanism of esterification is expected):**

(i) $H-O-H + Cl-\overset{\overset{O}{\|}}{C}-CH_3 \longrightarrow H-O-\overset{\overset{O}{\|}}{C}-CH_3 + HCl$

Acetic acid

(ii) $C_2H_5-O-H + Cl-\overset{\overset{O}{\|}}{C}-CH_3 \longrightarrow C_2H_5-O-\overset{\overset{O}{\|}}{C}-CH_3 + HCl$

Ethyl acetate

(iii) $C_6H_5\overset{\cdot\cdot}{O}H + \underset{Cl}{\overset{CH_3}{}}C=O \longrightarrow CH_3COOC_6H_5$

Phenyl acetate

2. **With ammonia:** When ammonia is passed in acetyl chloride, it gives acetamide.

$$CH_3COCl + NH_3 \longrightarrow CH_3CONH_2 + HCl$$

Acetyl chloride   Acetamide

Uses of Acetyl Chloride:

(a) As an acetylating agent.

(b) An important organic reagent.

(c) In the preparation of acetic anhydride.

(d) In the detection and estimation of alcoholic and amino groups.

1.6.2 Acid Anhydride Derivative

Acetic anhydride is a colourless, pungent smelling liquid, insoluble in water. It is not as reactive as acetyl chloride and does not fume in moist air. Like acetyl chloride, it reacts with alcohol, phenol and amine, though less reactive, normally acetylation using acetic anhydride is carried out in the presence of sodium acetate or concentrated sulphuric acid as a catalyst.

$$R-\overset{\overset{O}{\|}}{C}-\overset{\cdot\cdot}{\underset{\cdot\cdot}{O}}-\overset{\overset{O}{\|}}{C}-R$$

Acid anhydrides are named by adding the word *"anhydride"* after the IUPAC name of the acid.

Table 1.6

Structure	Common name	IUPAC name
CH_3CO—O—CH_3CO	Acetic anhydride	Ethanoic anhydride
CH_2CO—O—CH_2CO	Succinic anhydride	Butanedioic anhydride
(phthalic anhydride structure)	Phthalic anhydride	o-benzenedioic anhydride

Method of Formation:

It has been prepared

1.   By dehydration of acetic acid: When vapours of acetic acid is passed over phosphorus pentoxide (dehydrating agent), it gives acetic anhydride with the removal of H_2O molecule.

$$CH_3CO\boxed{OH} \ + \ CH_3COO\boxed{H} \xrightarrow[\Delta]{P_2O_5} \begin{array}{c} CH_3CO \\ CH_3CO \end{array}\!\!\!>\!O \ + \ H_2O$$

(Acetic anhydride)

Chemical Reactions:

1.   With water: Acid anhydrides are hydrolyzed with water to acids.

$$(CH_3CO)_2O + H_2O \longrightarrow 2CH_3COOH$$
Acetic anhydride   Acetic acid

2.   With alcohol: Acid anhydrides react with ethyl alcohol to produce esters.

$$(CH_3CO)_2O + C_2H_5OH \longrightarrow CH_3COOC_2H_5 + CH_3COOH$$
Acetic anhydride   Ethyl   Ethyl acetate   Acetic acid
                  alcohol

3.   With ammonia: Ammonia rapidly reacts with acetic anhydride to give acetamide.

$$CH_3-CO-NH_2 + CH_3COOH \text{ (Ammonolysis)}$$
(Acetamide)

Ammonia

Uses of Acetic Anhydride:

1. As an acetylating agent for the manufacture of dyes, cellulose acetate etc.

2. In the manufacture of aspirin and some drugs.

Exercises

(A) Short and Long Answer Questions:

1. Describe the general methods of preparation, physical and chemical properties of acyl chlorides.

2. Give the preparation, physical and chemical properties of acid anhydrides.

3. How can you prepare acetyl chloride from carboxylic acids ?

4. How is citric acid synthesized ? Describe important chemical properties of citric acid.

5. Give methods of preparation of following:

    (a) Monochloro acetic acid

    (b) Dichloro acetic acid

    (c) Trichloro acetic acid

6. Give methods of preparation of malic and citric acid.

7. Give any two methods of preparation of acryllic acid and cinnamic acid.

8. How will you prepare cinnamic acid by Perkin reaction ?

9. How is acetic anhydride prepared ?

10. What happens when ......

    (a) Succinic acid is treated with C_2H_5OH/H^+

    (b) Phthalic acid is heated with sodalime

    (c) Citric acid is heated at 425 K

(d)   Citric acid is treated with HI

(e)   Succinic acid is treated with sodium bicarbonate

(f)   Phthalic acid is treated with NH_3

(g)   Hydrolysis of acetyl chloride

(h)   Oxidation of cinnamic acid with $KMnO_4$

(i)   Bromination of cinnamic acid

(j)   Acryllic acid with C_2H_5OH/Na

(k)   Malic acid is treated with HI

11. What is the action of following on monochloro acetic acid?

(a) OH^- (b) I^-

(c) CN^- (d) NH_3

12. What is the action of following on cinnamic acid ?

(a) Br_2 (b) $KMnO_4$

(c) CrO_3

(B) Select the correct alternative from the following and rewrite the sentence:

1.   Which functional group is present in a carboxylic acid?
     (a) –COOH   (b) $-NO_2$
     (c) C–O–C   (d) –SH

2.   Which one of the following is a monocarboxylic acid ?
     (a) oxalic acid   (b)   succinic acid
     (c) formic acid   (d)   citric acid

3.   Derivatives of carboxylic acids are hydrolyzed into ......
     (a) alcohols   (b)   acyl chlorides
     (c) thioethers   **(d) carboxylic acids**

4.   Citric acid is ....
     (a) halo acid   **(b) hydroxy acid**
     (c) unsaturated acid   (d)   mineral acid

5. Which of the following compounds gives positive iodoform test ?

    (a) 3-hexanone   (b) 1-pentanol

    (c) acetone   (d) 3-pentanone

6. Acyl chloride can be obtained ....

    (a) by direct esterification   (b) by dehydration of acids

    (c) from cyanohydrin reaction **(d) from carboxylic acids**

7. On reaction with thionyl chloride ($SOCl_2$), acids are converted into ....

    (a) acid anhydrides   (b) alcohols

    (c) amines   **(d) acid chlorides**

8. Anhydrides can be converted into carboxylic acids by ....

    (a) oxidation   (b) ammonolysis

    (c) hydrolysis   (d) decarboxylation

9. Malic acid can be reduced with HI into ......

    (a) succinic acid   (b) tartaric acid

    (c) citric acid   (d) malonic acid

10. ...... is not a hydroxy acid.

    (a) Glycolic acid   (b) Malic acid

    (c) Succinic acid   (d) Citric acid

11. Phthalic acid on heating forms ....

    (a) phthalic anhydride   (b) benzoic acid

    (c) cinnamic acid   (d) succinic anhydride

12. ...... is a dicarboxylic acid.

    (a) Cinnamic acid   (b) Benzoic acid

    (c) Succinic acid   (d) Acrylic acid

13. ......is an unsaturated acid.

    (a) Phthalic acid   (b) Benzoic acid

    (c) Succinic acid   **(d) Acrylic acid**

14. Hydrolysis of acetic anhydride gives ....

    (a) acetic acid   (b) acetamide

    (c) ethyl acetate   (d) methyl acetate

15. ...... on reduction with sodium in ethanol gives propanoic acid.

    (a) Cinnamic acid   **(b) Acrylic acid**

    (c) Benzoic acid   (d) Phthalic acid

Chapter **2**...

Amines and Diazonium Salts

Contents ...

2.1 Introduction

Amines may be regarded as derivatives of ammonia, in which one or more hydrogen atoms have been replaced by alkyl groups. Amines like ammonia are weak bases ($K_b = 10^{-4}$ to 10^{-6}). This basicity is due to the unshared pair on the nitrogen atom.

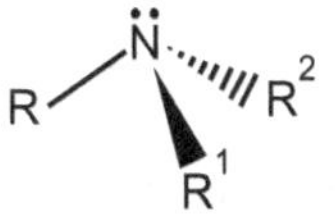

2.1.1 Classification

They are classified as primary, secondary or tertiary depending on whether one, two or all the three hydrogen atoms of ammonia molecule are replaced by alkyl groups.

| Ammonia | Primary amine | Secondary amine | Tertiary amine |

(2.1)

Secondary and tertiary amines may further be classified as simple or mixed depending on the fact that whether the alkyl groups are identical or different.

The characteristic groups in primary, secondary and tertiary amines are:

$$-NH_2 \qquad\qquad -\overset{|}{N}H \qquad\qquad -\overset{|}{\underset{|}{N}}-$$

Primary amine   Secondary amine   Tertiary amine

In addition to amines, tetra-alkyl derivatives of ammonium salts are also known and they are called quaternary ammonium salts e.g. $[(CH_3)_4N]^+Cl^-$.

2.1.2 Nomenclature

Amines are usually named after the alkyl groups attached to the nitrogen atom. Each name ends in the suffix 'amine'.

$$CH_3NH_2 \qquad\qquad C_2H_5NH_2 \qquad\qquad (CH_3)_2CH \cdot NH_2$$

Methyl amine   Ethyl amine   Isopropyl amine

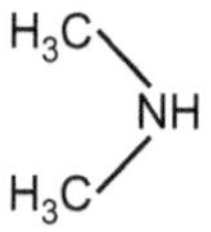

Dimethylamine 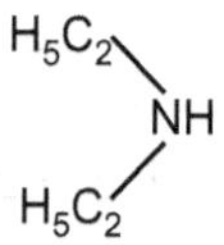   Diethylamine 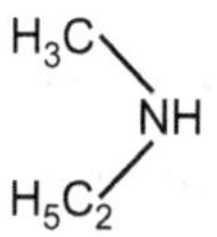   Methylethylamine

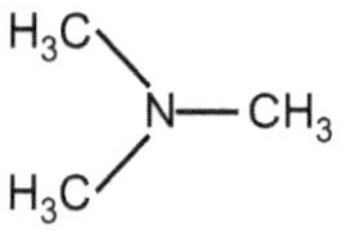

Trimethylamine 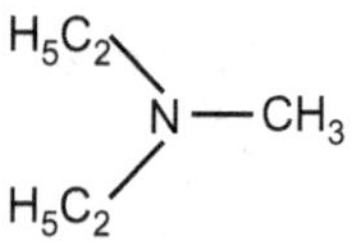   Methyldiethylamine 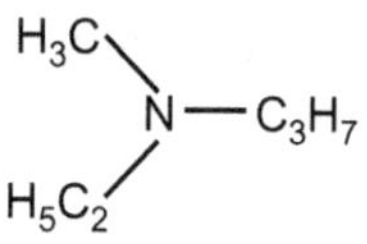   Methylethylpropylamine

According to IUPAC system of nomenclature, primary amines are named as amino alkanes, secondary amines as N-alkyl amino alkanes and tertiary amines as N, N-dialkyl amino alkanes e.g.

$$C_2H_5NH_2 \qquad\qquad H_3C-\overset{\overset{\displaystyle H}{|}}{N}-CH_3 \qquad\qquad H_3C-\overset{\overset{\displaystyle CH_3}{|}}{N}-CH_3$$

Amino ethane   N-methyl amino   N, N-dimethyl amino
                        methane   methane

IUPAC system is, however, adopted in cases where trivial system seems to be rather difficult e.g. The compound given below is 1-amino-2-methylhexane.

$$CH_3 \cdot CH_2 \cdot CH_2 \cdot CH_2 \cdot CH \cdot CH_2 \cdot NH_2$$
$$|$$
$$CH_3$$

2.1.3 Structure

Nitrogen has five valence electrons and so is trivalent with a lone pair. As per VSEPR theory, nitrogen present in amines is sp^3 hybridized and due to the presence of lone pair, it is pyramidal in shape instead of tetrahedral shape which is a general structure for most sp^3 hybridized molecules. Each of the three sp^3 hybridized orbitals of nitrogen overlap with orbitals of hydrogen or carbon depending upon the configuration of amines. Due to the presence of lone pair, the C–N–H angle in amines is less than 109° which is the characteristic angle of tetrahedral geometry. The angle of amines is near about 108°.

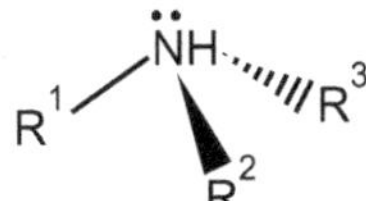

2.1.4 Methods of Preparation

Amines can be prepared by different methods as given below.

(a) From Alkyl Halide by Ammonolysis: When an aqueous or alcoholic solution of ammonia is heated with an alkyl halide at 100°C in a sealed tube, all the three types of amines along with quaternary ammonium salt are formed. The process is known as ammonolysis. If excess of ammonia is used, the main product is primary amine, while with excess of alkyl halide, the main product is tertiary amine.

$$C_2H_5I + H \cdot NH_2 \text{ (excess)} \longrightarrow C_2H_5NH_2 + HI$$

Ethyl amine

(Primary amine)

$$C_2H_5I \text{ (excess)} + H \cdot NH_2 \longrightarrow (C_2H_5)_3N + HI$$

Triethyl amine

(Tertiary amine)

(b) By Reduction of Nitriles or Cyanides: When reduced catalytically or with sodium and alcohol or lithium aluminium hydride, alkyl cyanides form primary amines.

$$H_3C - C \equiv N + 4H \xrightarrow{\text{LiAlH}_4} CH_3 - CH_2 - NH_2$$

Methyl cyanide   Ethyl amine

(c) From Unsubstituted Amides (Hoffmann degradation): This is a very convenient method for the preparation of primary amines. When acid amides containing upto six carbon atoms are treated with bromine and caustic potash, primary amines with one carbon atom less are formed.

$$RCONH_2 + Br_2 + 4KOH \longrightarrow RNH_2 + 2KBr + K_2CO_3 + 2H_2O$$

Amide   Primary amine

$$CH_3CONH_2 + Br_2 + 4KOH \longrightarrow CH_3NH_2 + 2KBr + K_2CO_3 + 2H_2O$$

Acetamide   Methyl amine

(d) By Gabriel synthesis (from Phthalimide): Phthalimide is first converted into potassium phthalimide by treatment with caustic potash. The potassium phthalimide, when heated with alkyl halide gives N-alkyl phthalimide, which on hydrolysis with 20 percent hydrochloric acid under pressure forms primary amine.

Phthalimide   Potassium phthalimide

R—NH$_2$ + Phthalic acid   Hydrolysis   N-alkyl phthalimide

Amine

2.1.5 Reactions

Different types of reactions has been carried out by amines.

1. Carbylamine reaction: When heated with chloroform and alcoholic caustic potash, primary amines form isocyanides having extremely disagreeable odour. This is a test for primary amines.

$C_2H_5NH_2 + CHCl_3 + 3KOH \longrightarrow C_2H_5N \rightleftharpoons C + 3KCl + 3H_2O$
Ethyl amine   Ethyl isocyanide

$C_6H_5NH_2 + CHCl_3 + 3KOH \longrightarrow C_6H_5N \rightleftharpoons C + 3KCl + 3H_2O$
Aniline   Phenyl isocyanide

2. Schotten-Baumann reaction: The amino groups of the amines are acylated when they are treated with aromatic acid chlorides or sulphonyl chloride in the presence of caustic soda solution, is known as Schotten-Baumann reaction. This technique fails in case of aliphatic acid chlorides as they are hydrolyzed quickly in aqueous alkali.

$$C_6H_5NH_2 + C_6H_5COCl \xrightarrow{KOH} C_6H_5NHCOC_6H_5 + HCl$$
Aniline   Benzanilide

$$C_6H_5NH_2 + C_6H_5SO_2Cl \xrightarrow{KOH} C_6H_5NHSO_2C_6H_5 + HCl$$
Aniline   Benzene   N-phenyl sulphonamide
            sulphonyl chloride

3. Electrophilic Substitution of Aniline: The presence of amino group has an activating effect on the nucleus with a consequent increase in the electron density at *ortho-* and *para-* positions. Thus the aromatic amines undergo typical electrophilic substitution at *ortho-* and *para-* positions.

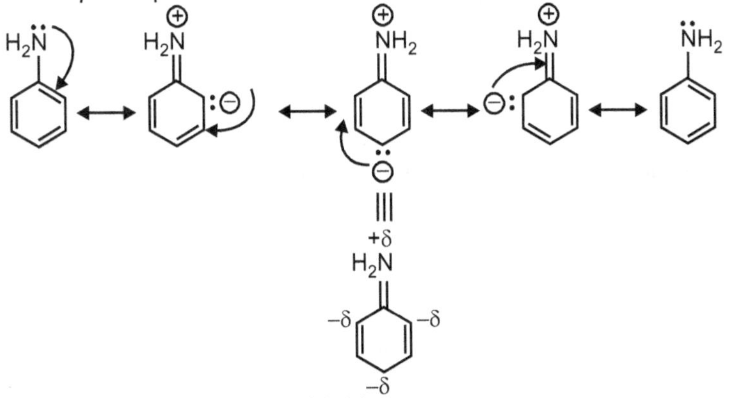

Hybrid structure

(A) Nitration: When nitric acid is used for nitration, the amines are oxidized giving various oxidation products. In order to prevent this, the amino group is protected, this amino group is protected by acetylation and anilide thus obtained is subjected to nitration. The nitroacetanilide thus obtained is subjected to hydrolysis to get the nitrated amine.

Aniline → (CH₃CO)₂O → Acetanilide → HNO₃/H₂SO₄ → o-Nitroacetanilide → HOH → o-Nitroaniline

It is observed that when acetic anhydride is used as an acetylating agent the product on nitration is mainly the o-nitroacetanilide whereas in case of acetic acid as a acetylating agent the nitration product is mainly the p-isomer.

Aniline → CH₃COOH → Acetanilide → HNO₃/H₂SO₄ → p-Nitroacetanilide → HOH → p-Nitroaniline

(B) Bromination: The amino group activates the nucleus to such an extent that the substitution takes place at two ortho- and one para- position. Thus the reaction with chlorine and bromine gives the trisubstitution product e.g. in reaction with bromine, aniline forms 2, 4, 6-tribromo aniline.

Aniline → Br₂ → 2, 4, 6-Tribromoaniline

If however, the amino group is acetylated, the activating influence of –NHCOCH₃, is much less as compared to –NH₂ group and in that case only monosubstitution product is obtained.

Aniline $\xrightarrow{CH_3COCl}$ Acetanilide $\xrightarrow{Br_2}$ p-Bromoacetanilide $\xrightarrow{HOH/H^+}$ p-Bromoaniline

(C) Sulphonation: The sulphonation takes place at *ortho-* and *para-* positions. For example, aniline is mixed with equimolecular quantity of sulphuric acid and resulting aniline acid sulphate is heated at $60°$ to give orthoanilic acid (o-amino benzene sulphonic acid) and at $180°$ gives sulphanilic acid (p-amino benzene sulphonic acid).

Aniline $\xrightarrow{H_2SO_4}$ Aniline acid sulphate $\xrightarrow{-H_2O}$ Phenylsulphamic acid $\xrightarrow[\Delta]{60}$ Orthoanilic acid

$\xrightarrow[]{\Delta \mid 180°}$ Sulphanilic acid

Sulphanilic acid have Zwitter ion structure:

$^-O_3S-\langle\bigcirc\rangle-\overset{+}{N}H_3$

2.2 Diazonium Salt

2.2.1 Introduction

These are the compounds which contain diazonium ion group $(-N^+ \equiv N)$ with anion X^- like Cl^-, Br^-, NO_2^-, HSO_4^-, BF_4^- attached to aryl unit is called as Diazonium salt.

Benzene diazonium ion

Benzene diazonium chloride

There are some examples of diazonium salts such as,

$\overset{+}{N_2}\overset{-}{Cl}$ (benzene ring)	Benzene diazonium chloride
O_2N—(benzene ring)—N_2Cl	p-nitro benzene diazonium chloride
(benzene ring)—N_2BH_4	Benzene diazonium fluoroborate
(benzene ring)—N_2HSO_4	Benzene diazonium hydrogen sulphate

Diazotization: The process of converting primary aromatic amine into its diazonium salt by treatment with nitrous acid at 273 K is known as diazotization.

e.g. Conversion of aniline into benzene diazonium chloride by reaction with nitrous acid ($NaNO_2$ + HCl) at 273 K.

Benzene Diazonium Chloride: The structure of benzene diazonium chloride is as follows.

$$\text{(benzene ring)}-\overset{+}{N_2}\overset{-}{Cl} \quad \text{or} \quad \text{(benzene ring)}-N{=}N{-}Cl$$

2.2.2 Preparation of Benzene Diazonium Chloride

Benzene diazonium chloride is obtained by treating aniline with nitrous acid at 0-5°C (ice bath). The HNO_2 is prepared in situ from sodium nitrate and HCl.

$$\text{(benzene ring)}-NH_2 + NaNO_2 \xrightarrow[273\text{-}278\ K]{HCl} \text{(benzene ring)}-\overset{+}{N_2}\overset{-}{Cl} + 2H_2O$$

Benzene diazonium chloride

The following steps take place in the preparation of benzene diazonium chloride as,

1. $\text{(benzene ring)}-NH_2 + HCl \xrightarrow{273\ K} \text{(benzene ring)}-\overset{+}{N}H_3\overset{-}{Cl}$

Benzene ammonium chloride

2. $NaNO_2 + HCl \xrightarrow{273\ K} HNO_2 + NaCl$

Nitrous acid

3. $\text{(benzene ring)}-\overset{+}{N}H_3\overset{-}{Cl} + HNO_2 \xrightarrow{273\ K} \text{(benzene ring)}-\overset{+}{N_2}\overset{-}{Cl} + 2H_2O$

Benzene diazonium chloride

2.2.3 Reactions

Benzene diazonium chloride is a highly reactive compound and it shows two types of reactions.

1. The reaction in which replacement of N_2Cl takes place.
2. The reaction in which two nitrogen atoms are retained.

(I) The reaction in which replacement of $-N_2Cl$ takes place:

(a) Replacement by Halogen and Cyanide (Sandmeyer's reaction): In Sandmeyer's reaction, benzene diazonium chloride is treated with cuprous chloride or bromide and cuprous cyanide in the presence of their corresponding acid to give chlorobenzene or bromobenzene and benzonitrile respectively.

Replacement by Halogen:

$$C_6H_5-N_2Cl + CuCl \xrightarrow{\text{HCl / 298 K}} C_6H_5-Cl + N_2 + CuCl$$

Chlorobenzene

$$C_6H_5-N_2Cl + CuBr \xrightarrow{\text{HBr / 298 K}} C_6H_5-Br + N_2 + CuCl$$

Bromobenzene

Replacement by Cyanide:

$$C_6H_5-N_2Cl + CuCN \xrightarrow{\text{KCN / 298 K}} C_6H_5-CN + N_2 + CuCl$$

Benzonitrile

(b) Replacement by Iodine: When benzene diazonium chloride is treated with aqueous KI, it gives iodobenzene.

$$C_6H_5-N_2Cl + KI \xrightarrow{\text{H}_2\text{O}} C_6H_5-I + N_2 + KCl$$

Iodobenzene

(c) Replacement by –OH: When aqueous solution of benzene diazonium chloride is slowly added to dilute and boiling H_2SO_4, the diazonium group is replaced by –OH group and phenol is formed.

$$C_6H_5-N_2Cl + H_2O \xrightarrow{\text{H}_2\text{SO}_4 \text{ / boil}} C_6H_5-OH + N_2 + HCl$$

Phenol

(II) The reaction in which two nitrogen atoms are retained:

Benzene diazonium chloride on reduction with aqueous sodium sulphite at 373 K gives phenyl hydrazine.

Reduction of benzene diazonium chloride:

$$C_6H_5-N_2Cl + Na_2SO_3 \xrightarrow[\text{2H}_2\text{O}]{\text{373 K}} C_6H_5-NH\text{-}NH_2 + Na_2SO_4 + HCl$$

Phenyl hydrazine

(III) Coupling Reaction:

The aromatic compounds like phenols, naphthols and anilides are treated with benzene diazonium chloride to form Ar-N=N–Ar. This form of compound is called as azo compound is and the formation of azo compound is called as ***coupling reaction***.

The coupling reaction takes place at 273-278 K, in coupling reaction, two nitrogen atoms are attached to the aromatic ring.

1. Synthesis of Methyl Orange: It takes place in two steps:

Step-I: Formation of Diazonium salt from Sulphanilic acid:

$$HO_3S-\!\!\bigcirc\!\!-NH_2 \xrightarrow[\text{273-278 K}]{\text{NaNO}_2\,/\,\text{HCl}} HO_3S-\!\!\bigcirc\!\!-N_2Cl$$

Sulphanilic acid   Diazonium salt

Step-II: Formation   of Methyl Orange:

Diazonium salt is coupled with N, N-dimethyl aniline in dilute HCl at 273 K to give methyl orange.

2.   Synthesis of Congo Red: It is a bis azo dye prepared from benzidine and naphthionic acid. It takes place in two steps.

$$HO_3S-\!\!\bigcirc\!\!-N_2Cl \quad + \quad \bigcirc\!\!-N\!\!<^{CH_3}_{CH_3}$$

Diazonium salt   N, N-Dimethyl aniline

dil. HCl

$$HO_3S-\!\!\bigcirc\!\!-N\!=\!N-\!\!\bigcirc\!\!-N\!\!<^{CH_3}_{CH_3}$$

Methyl Orange

Step-I : Formation of Tetrazotized Benzidine:

$$H_2N-\!\!\bigcirc\!\!-\!\!\bigcirc\!\!-NH_2 \xrightarrow[\text{273-278 K}]{\text{NaNO}_2\,/\,\text{HCl}} ClN_2-\!\!\bigcirc\!\!-\!\!\bigcirc\!\!-N_2Cl$$

Benzidine   Tetrazotized Benzidine

Step-II: Formation of Congo Red:

Naphthionic acid   +   Tetrazotized Benzidine   +   Naphthionic acid

–2HCl | Coupling

Congo Red

Exercises

(A) Write a short note on :

1. Sandmeyer's reaction
2. Coupling reaction
3. Hoffmann degradation
4. Congo red
5. Nitration of aniline
6. Sulphonation of aniline
7. Diazonium salt
8. Diazotization

(B) Short and Long Answer Questions :

1. Give the synthesis of congo red and methyl orange.
2. Explain coupling reaction with examples.
3. Define diazotization. How will you prepare it ?
4. Give any two methods for the preparation of primary amines.
5. Give classification and structure of amines.
6. Explain the electrophilic substitution reactions in aniline.
   (a) Nitration, (b) Sulphonation, (c) Bromination
7. Explain carbylamine reaction and Schotten-Baumann reaction.
8. How will you prepare primary amine by Gabriel and Hoffmann degradation method ?
9. How will you prepare iodobenzene, phenyl hydrazine and benzonitrile from benzene diazonium chloride ?

(C) Choose the correct alternative for each of the following and rewrite the sentence.

1. Which of the following is a 3° amine?
   (a) 1-methyl cyclohexylamine
   (b) triethylamine
   (c) tert-butylamine
   (d) N-methyl aniline
2. The source of nitrogen in Gabriel synthesis of amines is ....
   (a) Sodium azide, NaN_3
   (b) Sodium nitrite, $NaNO_2$
   (c) Potassium cyanide, KCN
   (d) Potassium phthalimide, $C_6H_4(CO)_2N^-K^+$
3. Which of the following cannot be prepared by Sandmeyer's reaction?
   (a) Chlorobenzene
   (b) Bromobenzene
   (c) Phenyl chloride
   (d) Fluorobenzene

4. What is the general formula of diazonium salt?
   (a) $RN_2^+X^-$ (b) RN^+
   (c) RXI   (d) $RN_2^+HSO^{-2}$

5. What is the known name of the given reaction?
   $$C_6H_5\overset{+}{N_2} + CuCl \longrightarrow C_6H_5Cl + N_2 + Cu^+$$
   (a) Gattermann's reaction
   (b) Sandmeyer's reaction
   (c) Dehydrogenation reaction
   (d) Esterification reaction

6. Which of the following amine will form stable diazonium salt at 273-283 K?
   (a) $C_6H_5NH_2$ (b) $C_6H_5N(CH_3)_2$
   (c) $C_2H_5NH_2$ (d) $C_6H_5CH_2NH_2$

7. What happens when benzene diazonium chloride is treated with potassium cyanide in the presence of copper powder ?
   (a) Benzophenone   (b) Methyl isocyanide
   (c) Acetonitrile   **(d) Benzonitrile**

8. By treating diazonium salts with cuprous cyanide or KCN and copper powder, it forms which of the following compound?
   (a) Citric acid   (b) Benzoic acid
   (c) Aryl nitrile   (d) Oxalic acid

9. Azo dye is prepared by coupling of phenol and which of the following compound?
   (a) Diazonium chloride   (b) o-nitro aniline
   (c) Benzoic acid   (d) Chlorobenzene

10. When diazonium salt solution is treated with KI, it forms which of the following compound?
    (a) Bromobenzene   **(b) Iodobenzene**
    (c) Phenol   (d) Acid

11. Diazonium salt is coupled with N, N-dimethyl aniline in dilute HCl at 273 K to give ....
    (a) Methyl Orange   (b) Iodobenzene
    (c) Benzene   (d) Congo Red

12. ...... is formed when aqueous solution of benzene diazonium chloride is slowly added to dilute and boiling H_2SO_4.
    (a) Benzene   **(b) Phenol**
    (c) Bromobenzene   (d) Iodobenzene

ANSWERS

1. (b)	2. (d)	3. (d)	4. (a)	5. (b)	6. (a)	7. (d)	8. (c)
9. (a)	10. (b)	11. (a)	12. (b)				

✱✱✱

Chapter **3**...

Carbohydrates

Contents ...

3.1 Introduction

Carbohydrates are biological molecules made of C, H, and O. This composition gives carbohydrates their name *Carbo-* (carbon) plus -*hydrate* (water). In this chapter, we will learn more about the chemistry of carbohydrates.

Carbohydrates perform numerous roles in living organisms. They serve for the storage of energy (e.g. Starch and Glycogen) and as structural components (e.g. Cellulose in plants and chitin in arthropods). The ribose is an important component of coenzymes (e.g. ATP, FAD and NAD) and the backbone of the genetic molecules (e.g. RNA and DNA).

Carbohydrates are found in a wide variety of natural and processed foods like cereals (wheat, maize, and rice), potatoes, bread, pizza or pasta. Sugars appear in human diet mainly as table sugar (sucrose, extracted from sugar cane or sugar beets) and lactose (abundant in milk). Glucose and fructose, both of which occur naturally in honey, many fruits, and some vegetables.

The Formula for Carbohydrates: Carbohydrates have usually H and O atom with a ratio of 2 : 1 (as in water) and have the empirical formula $C_nH_{2n}O_n$ or $C_n(H_2O)_n$ (hydrates of C), where n = number of atoms.

As the oxygen is present in the form of OH and aldehydes or ketones, carbohydrates are called *'polyhydroxy aldehydes (glucose) or ketones (fructose) or compounds that produce such substances upon hydrolysis.*

Carbohydrates are also called saccharides. The word saccharide comes from the Greek word *'sákkharon'* meaning "*sugar*".

3.2 Classification of Carbohydrates (Monosaccharides, Disaccharides, Oligosaccharides and Polysaccharides)

All carbohydrates are broadly classified into two groups: (I) Simple carbohydrates and (II) Complex carbohydrates.

They are subdivided into four categories: monosaccharides, disaccharides, oligosaccharides and polysaccharides (Refer Fig. 3.1).

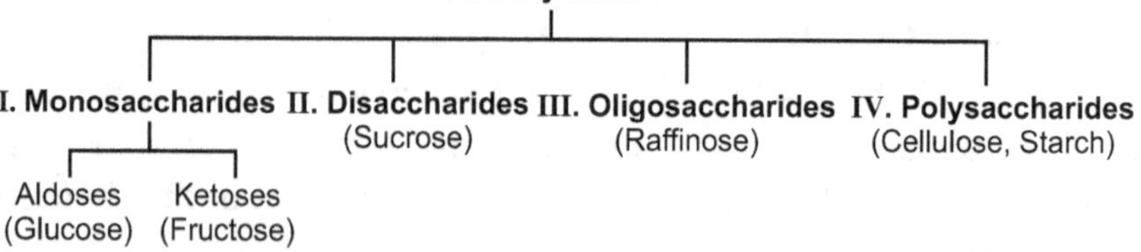

Fig. 3.1: Classification of carbohydrates

(I) Monosaccharides:

Monosaccharides are defined as polyhydroxy aldehydes or ketones, which cannot be further hydrolyzed to simple sugars.

Monosaccharides contain one sugar unit having general formula $(CH_2O)_n$. They typically contain three to seven carbon atoms (n = 3-7). The most common of which is glucose. Most of the oxygen atoms in monosaccharides are found in the form of hydroxyl groups (OH), but one of them is part of a carbonyl group (C=O).

Monosaccharides are again sub-divided into various groups according to their functional (carbonyl) groups and the number of carbon atoms present in the monosaccharides.

(A) Classification based on the nature of the carbonyl group:

According to the nature and the position of the carbonyl (C = O) group, monosaccharides can also be sub-categorized as follows:

(a) Aldoses: If the sugar has an aldehyde group, meaning that the CHO group is the last one in the chain, it is known as an aldose.

Examples: Glucose, Galactose, Mannose.

(b) Ketoses: If the sugar has ketone group, meaning that the C=O group is in the chain, it is known as ketose.

Examples: Fructose and Sorbose.

(B) Classification based on the number of carbon atoms:

Monosaccharides may be further sub-classified as follows depending on the number (n) of carbon atoms.

(i) Trioses: These have three carbon atoms per molecule having general formula $(C_3H_6O_3)$.

Examples: Aldotrioses (Glyceraldehyde) and Ketotrioses (Dihydroxy acetone).

(ii) Tetroses: These have four carbon atoms per molecule having general formula $(C_4H_8O_4)$.

Examples: Aldotetroses (Erythrose) and Ketotetroses (Erythrulose).

(iii) Pentoses: These have five carbon atoms per molecule having general formula $(C_5H_{10}O_5)$.

Examples: Aldopentoses (Ribose) and Ketopentoses (Ribulose).

(iv) Hexoses: These have six carbon atoms per molecule having general formula $(C_6H_{12}O_6)$.

Examples: Aldohexoses (Glucose) and Ketohexoses (Fructose).

(II) Disaccharides:

The carbohydrates which composed of two monosaccharide units (simple sugars) and are joined by a glycosidic linkage are called as disaccharides (also called a double sugar). Like monosaccharides, disaccharides are soluble in water.

$$\text{Disaccharides} \xrightarrow{\text{Hydrolysis}} \text{2 Monosaccharides}$$

Examples: Maltose: composed of two glucose units.

Lactose: composed of galactose and glucose units.

Sucrose: composed of glucose and fructose units.

(III) Oligosaccharides:

An oligosaccharide is a saccharide polymer containing a small number, typically 3-10 monosaccharides (simple sugars). An oligosaccharide is from the Greek *'oligos'*, "*a few*", and *'sacchar'*, "*sugar*". They are formed by joining 3-10 monosaccharide units by special type of bond called *'glycosidic bonds'* that can be hydrolyzed to yield 3-10 monosaccharide units.

$$\text{Oligosaccharides} \xrightarrow{\text{Hydrolysis}} \text{3-10 Monosaccharides}$$

They are further sub-classified into trisaccharides, tetrasaccharides etc. depending on the number of monosaccharide units present in it.

Example: Raffinose is a trisaccharide composed of galactose, glucose, and fructose.

(IV) Polysaccharides:

They are relatively complex carbohydrates. They are polymers made up of many monosaccharides joined together by glycosidic bonds. They are insoluble in water, and have no sweet taste.

$$\text{Polysaccharides} \xrightarrow{\text{Hydrolysis}} \text{n Monosaccharides}$$

Example: Cellulose: composed of many glucose units.

Starch: composed of many glucose units.

3.3 Reducing and Non-reducing Sugars

Sugars can also be divided into two groups depending on their chemical behaviors: reducing sugars and non-reducing sugars.

3.3.1 Reducing Sugars

'A reducing sugar is a carbohydrate that gives a positive test with Tollen's and Benedict's solution.

In this test a sugar containing aldehyde or ketone group can be oxidized into three different types of acidic sugars depending on the type of oxidizing agent used. Weak oxidizing agents such as Tollen's and Benedict's solutions oxidize the aldehyde into an aldonic acid.

Three "visual" tests for aldehydes that can only be shown by reducing sugars are as follows.

1. Fehling's Test:

This test involves the addition of Fehling's solution into the carbohydrate, and the mixture is heated for a brief period. 'In this test, a sugar containing aldehyde group changes the colour of a blue Cu(II) solution to red Cu(I) [CuO]'.

This test was developed by German chemist Hermann von Fehling in 1849.

Fehling's solution: It is a chemical reagent used to differentiate reducing and non-reducing sugars. Fehling's solution is prepared by combining two separate solutions, known as Fehling's A and Fehling's B. Fehling's A is aqueous solution of copper (II) sulfate, which is deep blue in colour. Fehling's B is a colourless solution of aqueous potassium sodium tartarate (also known as Rochelle salt) made in a sodium hydroxide. Both solutions A and B are prepared separately and stored in rubber stoppered bottles. Whenever required, the Fehling's reagent is prepared fresh by mixing equal volumes of solution A and B.

The purpose behind using the tartarate is that it coordinates to the copper (II) and helps prevent it from crushing out of the solution.

$$\text{Aldehyde} + Cu^{2+} \longrightarrow \text{Aldonic acid} + Cu_2O$$

Aldehyde   Ditartarate complex   Red ppt
(Reducing sugar)   (Fehling solution)

2. Benedict's Test:

This test involves the addition of Benedict's solution into the carbohydrate, and the mixture is heated for a brief period. A positive test with Benedict's reagent is shown by a colour change from clear blue to a brick-red precipitate.

Benedict's solution: It is a slightly modified version of Fehling's solution.

Fehling's solution uses citrate while Benedict's solution uses tartarate, which provides better stability for the copper (II). Like Fehling's solution, it is best made fresh. The ingredients are copper (II) sulphate, sodium carbonate, and sodium citrate.

$$\underset{\substack{\text{Aldehyde} \\ \text{(Reducing sugar)}}}{R-\overset{O}{\underset{H}{\|}}} \quad + \quad \underset{\substack{\text{Citrate complex} \\ \text{(Benedict's solution)}}}{Cu^{2+}} \quad \longrightarrow \quad \underset{}{R-\overset{O}{\underset{OH}{\|}}} \quad + \quad \underset{\text{Red ppt}}{Cu_2O}$$

In addition, keto sugars also give a positive test for Benedict's solution because keto sugars could be converted to aldehyde sugars through the enediol intermediate under the reaction conditions. Therefore, all monosaccharides both aldoses and ketoses show a positive behavior in the Benedict's test and considered as reducing sugars.

Example: D-fructose, which is a keto sugar (ketose) will give a positive test for Benedict's solution because of the ability of ketoses to get converted to aldoses (aldehydes).

Aldose D-glucose	Enediol	Ketose D-fructose

Benedict's test has been used in the old days to detect excess blood sugar in diabetic patients. This test shows positive behavior for all reducing sugars which include maltose and lactose and therefore not a very good test for glucose in the urine.

3. Tollen's Test:

In this test, an aldehyde oxidation result in a beautiful "mirror" of silver metal to precipitate on the reaction vessel. It was named after its discoverer, the German chemist Bernhard Tollens.

Tollen's reagent: The active ingredient in the Tollen's test, $[Ag(NH_3)_2]^+$, does not have a long shelf life and like the Fehlings' and Benedict solutions is best prepared fresh. The reagent consists of a solution of silver nitrate, ammonia and some sodium hydroxide (to maintain a basic pH of the reagent solution). Then, addition of aqueous ammonia (NH_3) results in the formation of the silver-ammonia complex which is the active oxidant.

The sample to be tested is then added to the freshly prepared reagent solution. A positive test results in a beautiful mirror of silver metal being precipitated out on the reaction vessel.

$$\text{Aldehyde} + 2Ag^+ + 2OH^- \longrightarrow \text{(carboxylic acid)} + Ag$$

Aldehyde
(Reducing sugar)   Silver mirror

The most important thing to note is that all of above tests can be carried out in basic solution. Because acidic conditions might hydrolyze any acetals present to hemiacetals, giving a false positive test. In addition the base considerably speeds up the rate of ring-chain tautomerism (i.e. interconversion between the cyclic hemiacetal form and the linear aldehyde form). The bottom line here is that adding base has the effect of increasing the concentration of the starting aldehyde.

3.3.2 Non-reducing Sugars

A non-reducing sugar is a carbohydrate that is not oxidized by a weak oxidizing agent (an oxidizing agent that oxidizes aldehydes but not alcohols, such as the Tollen's reagent) in basic aqueous solution. The characteristic property of non-reducing sugars is that, in basic aqueous medium, they do not generate any compounds containing an aldehyde group. As non-reducing sugars do not have the aldehyde group, they cannot reduce copper (II) (blue) to the copper (I) (red).

Example: Sucrose is the most common non-reducing sugar. The linkage between the glucose and fructose units in sucrose, which involves aldehyde and ketone groups, is responsible for the inability of sucrose to act as a reducing sugar. It proves that it neither contains any hemi-acetal group nor a hemi-ketal group. That is the reason it is quite stable in water.

Glucose   Fructose

α, β (1$\rightarrow$2) glycosidic bond

The C_1 (anomeric) carbon of glucose has a glycosidic link to the C_2 of fructose so that neither can open to its straight-chain form and neither can form a free aldehyde (although fructose is described as a ketose, it can form the aldehyde group via tautomerisation).

Table 3.1: Difference between reducing and non-reducing sugars

Sr. No.	Reducing Sugars	Non-reducing Sugars
1.	Such sugar bears a free aldehyde (–CHO) or ketone (–CO) group.	Such sugar does not bear a free aldehyde (–CHO) or ketone (–CO) group.
2.	These type of sugars are basically an acetal (in its ring form) at the anomeric carbon.	These type of sugars are basically no acetal at the anomeric carbon.
3.	Reducing sugars have the capacity to reduce cupric ions of Benedict's/Fehling's solution to cuprous ions.	Reducing sugars fail to reduce cupric ions of Benedict's/ Fehling's solution to cuprous ions.
4.	Reducing sugars give a positive reaction towards the Fehling's test.	Non-reducing sugars give a negative reaction towards the Fehling's test.
5.	They act as good reducing agent.	They do not act as good reducing agent.
6.	Examples: Include all monosaccharides and some disaccharides. Glucose, Fructose, Galactose, Ribose, Xylose, Mannose, Maltose and Lactose.	Examples: Include some disaccharides and all polysaccharides, Sucrose and Raffinose, Starch, Cellulose, and Glycogen.

3.4 General Properties of D-(+)-Glucose

Glucose is a white crystalline compound extremely soluble in water, but sparingly soluble in alcohol. It exists in anhydrous form (M.P. = 419 K) and monohydrate form (M.P. = 359 K).

There are three types of active groups in glucose; glycosidic –OH, alcoholic OH and aldehydic –CHO group, that are responsible for the following chemical properties.

1. Acylation reactions (Ester formation):

Glucose can form esters with carboxylic acids due to the presence of OH groups. Glucose reacts with five molecules of acetic anhydride $((CH_3CO)_2O)$ to form penta acetate derivative. It obviously indicates that the glucose contains five OH groups.

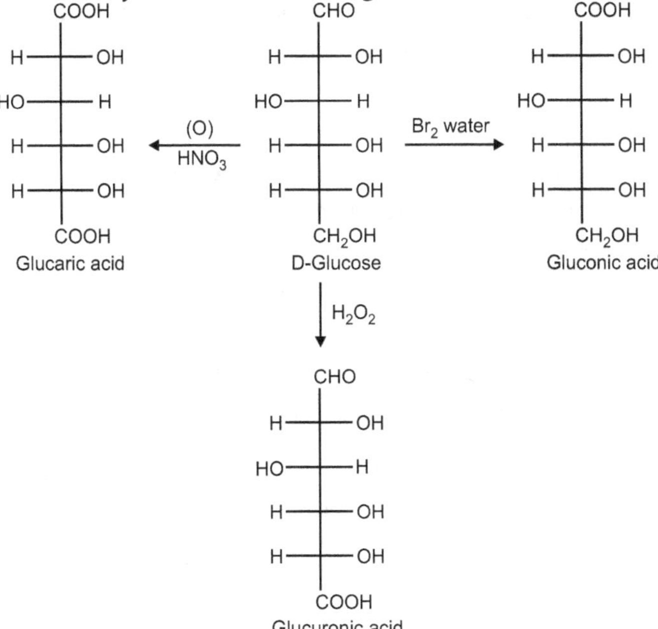

2. Oxidation reactions:

(i) Oxidation with Br water: Glucose when treated with bromine water, forms gluconic acid. The aldehyde group is oxidized to carboxylic group. First, bromine forms hypobromous acid (HOBr), with water and oxidizes the glucose to gluconic acid.

$$Br_2 + H_2O \rightleftharpoons HOBr + HBr$$

(ii) Oxidation with Nitric acid: Nitric acid oxidizes both the aldehyde and the terminal CH_2OH group of an aldose to COOH. The products of such oxidations are called aldaric (saccharic) acid.

(iii) Oxidation with hydrogen peroxide (H_2O_2): When glucose is oxidized with hydrogen peroxide (H_2O_2), glucuronic acid is formed. In this reaction only primary alcohol is converted into carboxylic group, whereas the aldehyde remains unchanged.

3. Reduction reaction:

Monosaccharides can be reduced by various reducing agents such as sodium-amalgam or by hydrogen under high pressure in the presence of catalysts. The reduction is due to the presence of aldehyde or ketone group. On reduction they yield alcohols.

When glucose is reduced by sodium amalgam, sorbitol is formed. Mannose yields mannitol and fructose yields a mixture of sorbitol and mannitol.

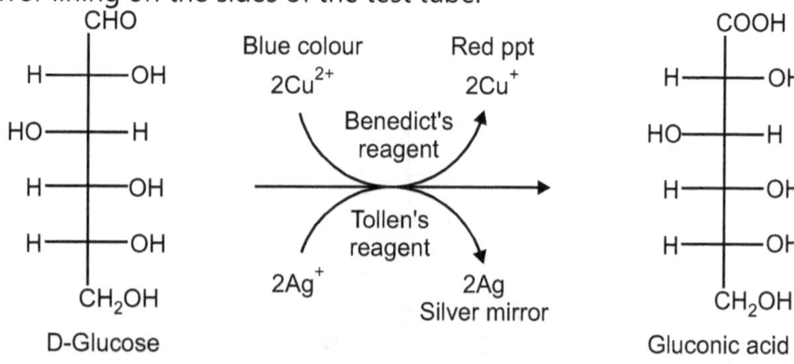

4. Reducing property:

Monosaccharides act as best reducing agents. They readily reduce oxidizing agents such as ferric cyanide, H_2O_2 and cupric ion. In such reactions, the sugar is oxidized at the carbonyl group and the oxidizing agent becomes reduced.

Glucose reduces Tollen's reagent, Fehling's reagent, Benedict's reagent etc. At the same time, glucose is oxidized to gluconic acid.

A standard test for the presence of reducing sugar is the reduction of Ag^+ in ammonia solution (Tollen's reagent) to yield a metallic silver mirror lining on the sides of the test tube.

5. Reaction with concentrated H_2SO_4:

Glucose is treated with concentrated H_2SO_4 or HCl and forms 5-hydroxy methyl furfural which on further heating yields levulinic acid and formic acid. This reaction is the basis of the colour test, known as 'Molish Test' for sugars.

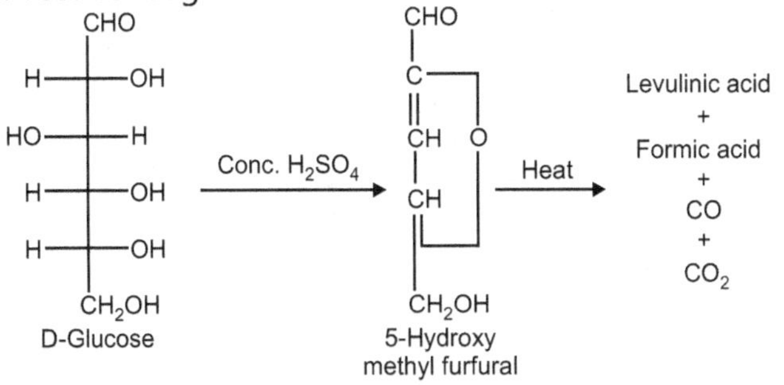

D-Glucose $\xrightarrow{\text{Conc. } H_2SO_4}$ 5-Hydroxy methyl furfural $\xrightarrow{\text{Heat}}$ Levulinic acid + Formic acid + CO + CO_2

6. Reaction with alanine:

The aldehyde group of glucose condenses with the amino group of alanine to form Schiff's base. Fructose also gives a Schiff's base with alanine.

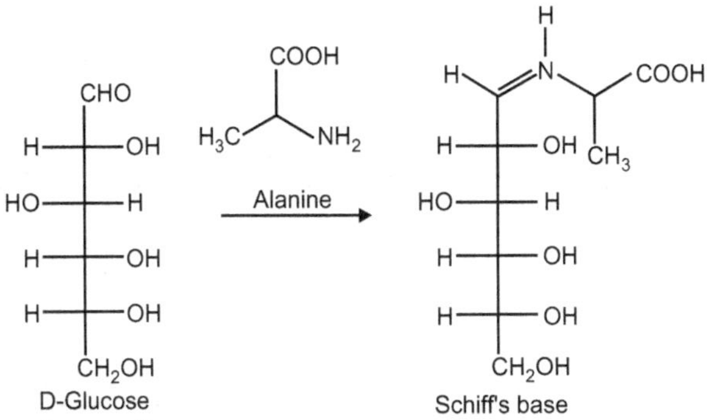

D-Glucose $\xrightarrow{\text{Alanine}}$ Schiff's base

7. Osazone formation:

One molecule of glucose condenses with one molecule of phenyl hydrazine to form soluble glucose phenylhydrazone, which on reaction with excess of phenylhydrazine gives yellow coloured crystals of phenyl glucosazone.

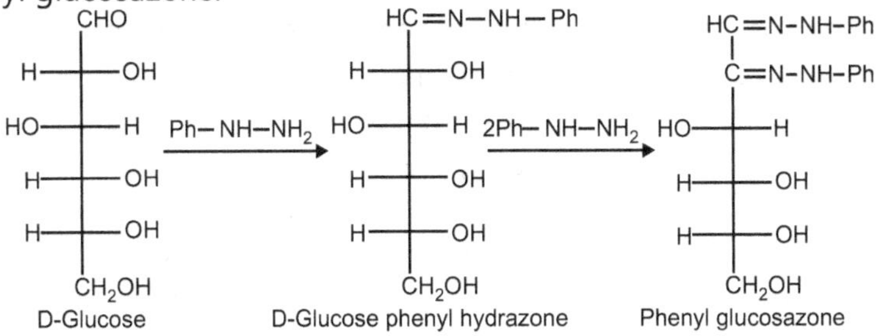

D-Glucose $\xrightarrow{\text{Ph}-NH-NH_2}$ D-Glucose phenyl hydrazone $\xrightarrow{2\text{Ph}-NH-NH_2}$ Phenyl glucosazone

The reaction of phenylhydrazine with fructose is similar to glucose. Here again three molecules of phenylhydrazine take part in the reaction. Fructose gives fructosazone.

8. Fermentation:

Fermentation is the process of converting a larger complex molecule into simple molecules by means of enzymes in an anaerobic condition. Glucose on fermentation by zymaze forms alcohol and CO_2.

$$\text{Glucose} \xrightarrow{\text{Zymase}} 2C_2H_5OH + CO_2$$
$$\text{Ethyl alcohol}$$

9. Epimerization:

The process by which one epimer is converted to another is called as epimerization.

Two sugars, which differ from one another only in configuration around a single carbon atom are termed "epimers". Glucose and mannose are epimers in respect of C_2.

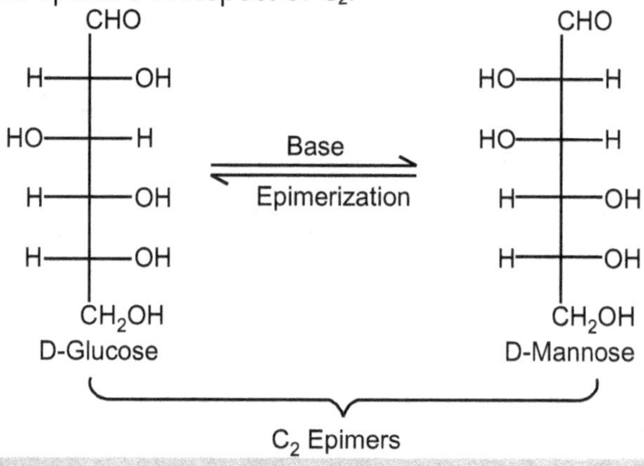

D-Glucose $\xrightarrow[\text{Epimerization}]{\text{Base}}$ D-Mannose

C_2 Epimers

3.5 Structure of D-(+)-Glucose

Glucose, also called dextrose is a simple sugar (monosaccharide) having the molecular formula $C_6H_{12}O_6$. Word glucose is derived from Greek glukos; meaning "sweet". Since, glucose is dextrorotatory (+), it is also called as 'dextrose.' Glucose contains six carbon atoms and an aldehyde group and is therefore referred to as an aldohexose.

It is found in fruits and honey and is the major sugar circulating in the blood of higher animals. It is the source of energy in cell function, and the regulation of its metabolism. Molecules of starch, the major energy-reserve carbohydrate of plants, consist molecules of thousands of linear glucose units. Another major compound composed of glucose is cellulose, which is also linear structure.

The glucose molecule can exist in an open-chain (acyclic) and ring (cyclic) form. The cyclic form resulted by intramolecular reaction between the aldehyde C_1 atom and the C_5 hydroxyl group to form an intramolecular hemiacetal. It is important to note that the linear form of glucose makes up less than 0.02% of the glucose molecules in water. The rest is one of two cyclic forms of glucose that are formed when the hydroxyl group on carbon 5 (C_5) bonds to the aldehyde carbon 1 (C_1).

Glucose is usually present in solid form as a monohydrate with a closed pyran ring (dextrose hydrate). In aqueous solution, on the other hand, it is an open chain to a small extent and is present in cyclic form predominantly as α- or β-pyranose, which partially mutually merge by mutarotation.

Thus, glucose can be represented by following different formulae (Refer Fig. 3.2).

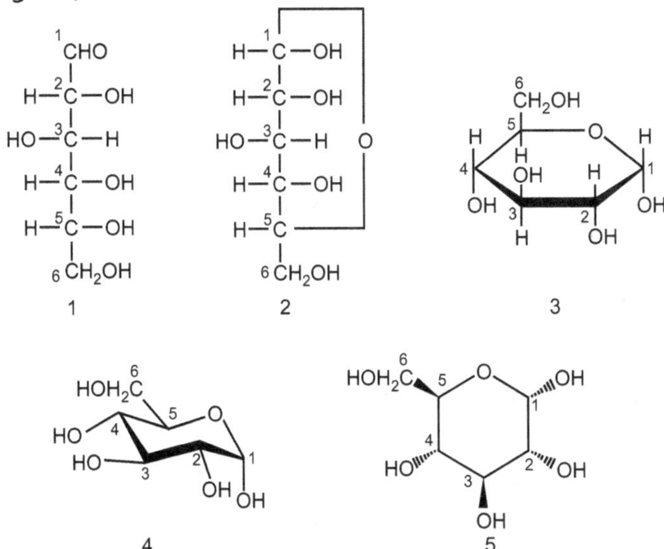

Fig. 3.2: Different forms and projections of D-Glucose (α-D-Glucopyranose) (1) Open chain, (2) Fischer projection, (3) Haworth projection, (4) Chair conformation, (5) Stereochemical view

(A) Open chain structure of glucose:

In open chain structure of glucose $(CHO)-(CHOH)_5-CH_2OH$, five hydroxyl (OH) groups are arranged in a specific way along its six-carbon with a terminal aldehyde group, shown as follows.

In this C_1 is part of an aldehyde group (CHO) and each of the other five carbons bears one hydroxyl group (–OH). The remaining bonds of the back bone carbons are satisfied by hydrogen atoms (–H).

The structure of glucose is described by following methods and reactions:

1. Molecular formula: Elemental analysis and molecular weight determination of glucose have led to the molecular formula to be $C_6H_{12}O_6$.

2. Presence of linear C-chain: Glucose on reduction with red phosphorus and hydroiodic acid (HI) gives n-hexane indicating the glucose contains six carbons in linear chain (C-C-C-C-C-C).

3. Presence of 5 OH groups:

(i) Glucose on reaction with acetic anhydride forms glucose pentaacetate indicating that it contains five hydroxyl (–OH) groups. In addition, it does not undergo dehydration easily, indicating that all five –OH groups are attached at five different carbons. Thus $CHO-C(OH)-C(OH)-C(OH)-C(OH)-CH_2OH$ chain is present in the glucose.

(ii) Presence of CH_2OH group: Glucose on reaction with the strong oxidizing agent (dil. HNO_3) forms dicarboxylic acid called saccharic/glucaric acid containing equal number of carbon atoms. It implies that the second carboxylic group in glucaric acid is derived by the oxidation of CH_2OH group attached at the end of chain. Thus, $CHO-C-C-C-C-CH_2OH$ chain is present in the glucose.

4. Presence and nature of carbonyl groups: Glucose gives typical reactions of carbonyl groups with different reagents.

    (i) Presence of carbonyl group: Glucose forms oxime with hydroxylamine, cyanohydrine with HCN, osazone with phenyl hydrazines indicating that glucose contains carbonyl group.

    (ii) Presence of CHO group: The presence of formyl group at the end of glucose is supported by following reactions.

        (a) The presence of formyl group: Glucose on reaction with the mild oxidizing agent (bromine water, Tollen's reagent) forms gluconic acid containing the same number of carbon atoms, indicating that carbonyl group is –CHO (formyl).

(b) Presence of formyl group at the end of glucose: Hydrolysis of glucose cyanohydrin forms a hydroxyl carboxylic acid containing one more carbon atom than glucose. The reduction of hydroxyl carboxylic acid with HI acid forms *n*-heptanoic acid. This indicates that formyl group (–CHO) in glucose is at the end of the linear chain. Thus CHO–C–C–C–C–CH$_2$OH chain is present in the glucose.

All above evidences show that the D-glucose is pentahydroxy aldehyde with linear structure as shown in structure (I).

Reactions:

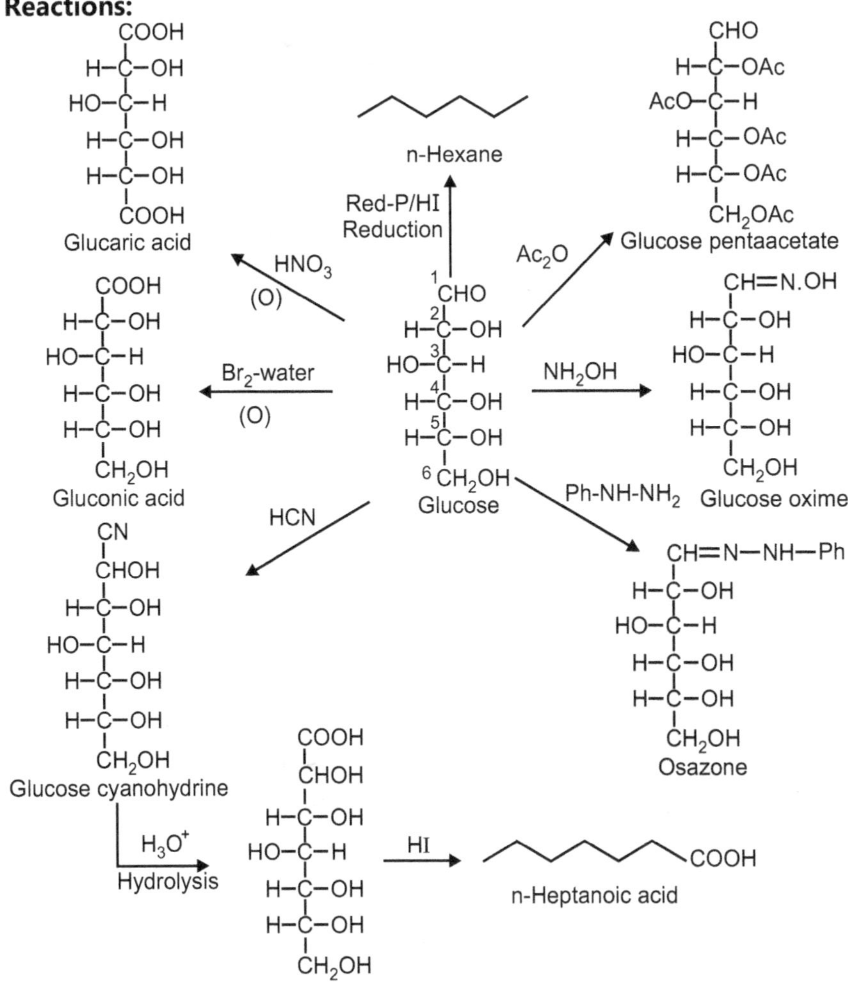

(B) The Configuration of D-Glucose (The Stereochemistry of D-Glucose):

The carbons labelled with an asterisk (*) in structure **I** are chiral. Thus there are 2^4 or sixteen (eight pairs of enantiomers), possible configurational isomers (8 are D-sugars (**I-VIII**) and 8 are L-sugars). All are known, some occur naturally and the others have been synthesized. The initial challenge was to determine which of the eight corresponded to glucose. This challenge was accepted and met in 1891 by the German chemist Emil Fischer, for which he was awarded the Nobel Prize in chemistry in 1902.

Glucose

Number of chiral centres = n = 4

Number of stereoisomers = $2^n = 2^4 = 16$ (8 pairs)

$$
\begin{array}{cccc}
\text{CHO} & \text{CHO} & \text{CHO} & \text{CHO} \\
\text{H}-\text{C}-\text{OH} & \text{H}-\text{C}-\text{OH} & \text{HO}-\text{C}-\text{H} & \text{HO}-\text{C}-\text{H} \\
\text{H}-\text{C}-\text{OH} & \text{HO}-\text{C}-\text{H} & \text{HO}-\text{C}-\text{H} & \text{H}-\text{C}-\text{OH} \\
\text{H}-\text{C}-\text{OH} & \text{H}-\text{C}-\text{OH} & \text{H}-\text{C}-\text{OH} & \text{H}-\text{C}-\text{OH} \\
\text{H}-\text{C}-\text{OH} & \text{H}-\text{C}-\text{OH} & \text{H}-\text{C}-\text{OH} & \text{H}-\text{C}-\text{OH} \\
\text{CH}_2\text{OH} & \text{CH}_2\text{OH} & \text{CH}_2\text{OH} & \text{CH}_2\text{OH} \\
\text{I} & \text{II} & \text{III} & \text{IV}
\end{array}
$$

$$
\begin{array}{cccc}
\text{CHO} & \text{CHO} & \text{CHO} & \text{CHO} \\
\text{H}-\text{C}-\text{OH} & \text{HO}-\text{C}-\text{H} & \text{H}-\text{C}-\text{OH} & \text{HO}-\text{C}-\text{H} \\
\text{H}-\text{C}-\text{OH} & \text{H}-\text{C}-\text{OH} & \text{HO}-\text{C}-\text{H} & \text{HO}-\text{C}-\text{H} \\
\text{HO}-\text{C}-\text{H} & \text{HO}-\text{C}-\text{H} & \text{HO}-\text{C}-\text{H} & \text{HO}-\text{C}-\text{H} \\
\text{H}-\text{C}-\text{OH} & \text{H}-\text{C}-\text{OH} & \text{H}-\text{C}-\text{OH} & \text{H}-\text{C}-\text{OH} \\
\text{CH}_2\text{OH} & \text{CH}_2\text{OH} & \text{CH}_2\text{OH} & \text{CH}_2\text{OH} \\
\text{V} & \text{VI} & \text{VII} & \text{VIII}
\end{array}
$$

The question is "which of the above structure describes the absolute configuration of glucose? How did Fischer determine which of the eight structures above was glucose?

The configuration of D-glucose is established by two different methods: (i) Elimination method, (ii) Method based on the configuration of lower sugar.

Method based on the configuration of lower sugar: Here we determine configuration of D-glucose based on the configuration of lower sugar (D-arabinose).

$$
\begin{array}{c}
CHO \\
HO-C-H \\
H-C-OH \\
H-C-OH \\
CH_2OH
\end{array}
$$

D-Arabinose

1.   This can be established with the help of Killiani-Fischer synthesis. When D-arabinose is subjected to Killiani Fisher synthesis a mixture of aldohexoses (**II**) and (**III**) are produced. One of them is D-Glucose and the other is D-Mannose.

D-Arabinose + HCN →

Upper path — Cyanohydrin (H_3O^+) → Aldonic acid (Heat) → Lactone (Na-Hg, H_3O^-) → **II**

$$
\begin{array}{cccc}
CN & COOH & & CHO \\
H-C-OH & H-C-OH & H-C-OH & H-C-OH \\
HO-C-H & HO-C-H & HO-C-H & HO-C-H \\
H-C-OH & H-C-OH & H-C & H-C-OH \\
H-C-OH & H-C-OH & H-C-OH & H-C-OH \\
CH_2OH & CH_2OH & CH_2OH & CH_2OH
\end{array}
$$

Lower path — Cyanohydrin (H_3O^+) → Aldonic acid (Heat) → Lactone (Na-Hg, H_3O^-) → **III**

$$
\begin{array}{cccc}
CN & COOH & & CHO \\
HO-C-H & HO-C-H & HO-C-H & HO-C-H \\
HO-C-H & HO-C-H & HO-C-H & HO-C-H \\
H-C-OH & H-C-OH & H-C & H-C-OH \\
H-C-OH & H-C-OH & H-C-OH & H-C-OH \\
CH_2OH & CH_2OH & CH_2OH & CH_2OH
\end{array}
$$

Cyanohydrin   Aldonic acid   Lactone   III

Thus D-Glucose and D-Mannose have the same configuration at C_3, C_4 and C_5 as that in D-Arabinose.

The question is "which of the two structures describes the absolute configuration of glucose?

2.   Killiani synthesis is used to decide the final configuration of glucose (**II/III**). Here, **II/III** is subjected to Killiani-Fischer synthesis, two aldoheptoses are obtained, which on oxidation forms two dibasic acids.

When **II** is subjected to Killiani synthesis-oxidation one is acid optically active while other is optically inactive, while **III** is subjected to Killiani synthesis-oxidation both are optically active.

Structure **II**:

```
        CHO                      COOH              COOH
         |                        |                 |
     H−C−OH                   H−C−OH            HO−C−H
         |                        |                 |
    HO−C−H    1. Killiani     H−C−OH            H−C−OH
         |       synthesis        |                 |
     H−C−OH    2. Oxidation   HO−C−H      +     HO−C−H
         |          →             |                 |
     H−C−OH                   H−C−OH            H−C−OH
         |                        |                 |
     CH₂OH                    H−C−OH            H−C−OH
         II                       |                 |
                               COOH              COOH
                           Dibasic acid       Dibasic acid
                         (Optically inactive) (Optically active)
```

Structure **III**:

```
        CHO                      COOH              COOH
         |                        |                 |
    HO−C−H                    H−C−OH            HO−C−H
         |                        |                 |
    HO−C−H    1. Killiani     HO−C−H            HO−C−H
         |       synthesis        |                 |
     H−C−OH    2. Oxidation   HO−C−H      +     HO−C−H
         |          →             |                 |
     H−C−OH                   H−C−OH            H−C−OH
         |                        |                 |
     CH₂OH                    H−C−OH            H−C−OH
        III                       |                 |
                               COOH              COOH
                           Dibasic acid       Dibasic acid
                         (Optically active) (Optically active)
```

When D-glucose is subjected to Killiani-Fischer synthesis, two aldoheptoses are obtained, which on oxidation forms two dibasic acids; one is optically active while other is optically inactive. Hence correct structure of glucose is **II** and therefore natural glucose is specifically D-glucose.

```
        CHO
         |
     H−C−OH
         |
    HO−C−H
         |
     H−C−OH
         |
     H−C−OH
         |
     CH₂OH
         II
   D-(+)-Glucose
```

(C) Objections against open chain structure of D-glucose:

The open chain structure of D-glucose has following objections:

1. The IR spectrum of glucose does not show the IR absorption band at 1720 cm^{-1}, which is a characteristic property of aldehyde.

2. The ^{1}H-NMR spectra of glucose do not show NMR signal at $\delta = 9.7$ ppm for aldehyde group.

3. Glucose does not undergo certain characteristic reactions of aldehyde:

(i)   Glucose does not form adduct with sodium bisulphite ($NaHSO_3$) or ammonia (NH_3) or 2,4-DNP.

(ii)   Glucose does not show Schiff's test.

(iii)   The glucose react with NH_2OH to form oxime but glucose penta acetate does not, this implies that aldehyde group is absent in glucose penta acetate.

(iv)   Glucose forms two isomeric methyl glucosides: Aldehydes normally react with two molecules of methanol to form acetal. But D-glucose when treated with methanol in the presence of dry HCl reacts with only one molecule of methanol to form two different methyl-D-glucosides (α and β). This indicates that CHO is absent in D-glucose.

Aldehyde   (dry HCl, CH_3OH, First molecule of methanol) → Hemiacetal   (dry HCl, CH_3OH, Second molecule of methanol) → Acetal

$C_6H_{12}O_6$ (D-Glucose)   (dry HCl, CH_3OH, First molecule of methanol) →

- $(C_6H_{11}O_5)$-OCH_3 α-**D-methyl glucoside**
- $(C_6H_{11}O_5)$-OCH_3 β-**D-methyl glucoside**

4.   Both α-D-glucose and β-D-glucose undergo mutarotation; a property known to be shown by γ-lactone.

(D) Cyclic/Ring structure of D-glucose:

These objections indicated that glucose have some other alternative structure.

Monosaccharides do not have a free aldehyde but instead of cyclic hemiacetal structure. Since monosaccharaides contain number of –OH and aldehyde group therefore any of the –OH (C_4/C_5) may combine with aldehyde group (C_1) to form intramolecular hemiacetal. As a result, monosaccharides has cyclic structure with 5 (furanose: similar to furan) or 6 (pyranose: similar to pyran) membered cyclic containing O atom.

Thus in case D-glucose during hemiacetal formation, C_5-OH combines with C_1 (it becomes chiral: anomeric carbon) and has thus, two possible arrangements of –OH and –H groups around it. In other words, D-glucose exists in two stereo isomeric forms; i.e. α-D-glucose and β-D-glucose with γ-lactone/1,4 oxide/ pyranose/6-membered ring structure.

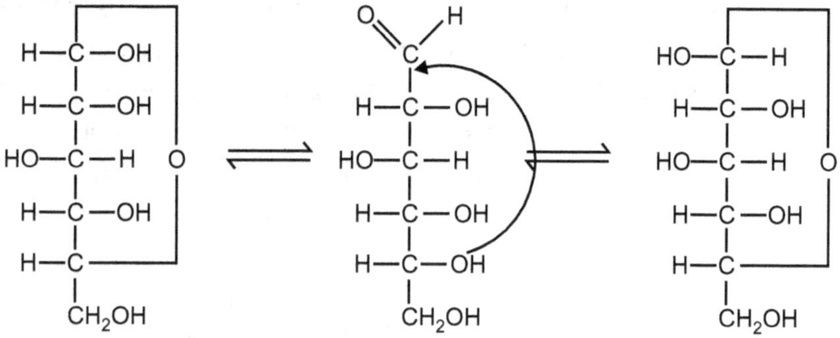

α-D-Glucose   **Open chain**   **β-D-Glucose**

This is also best represented by Haworth projection formula.

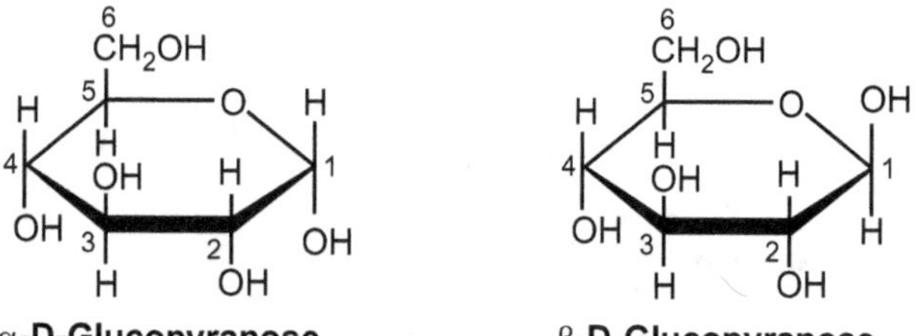

α-D-Glucopyranose   **β-D-Glucopyranose**
(α-D-Glucose)   **(β-D-Glucose)**

3.6 Epimers, Anomers and Mutarotation

3.6.1 Epimers

Enolization of an aldose challenges the stereochemistry at C_2. This process is called epimerization. Thus the process by which one epimer is converted to other epimer is called as epimerization.

Diastereomers that differ in stereochemistry at only one of their stereogenic centers are called epimers.

Examples: D-Glucose and D-mannose are epimers in respect of C_2, while D-Glucose and D-Galactose are epimers in respect of C_4.

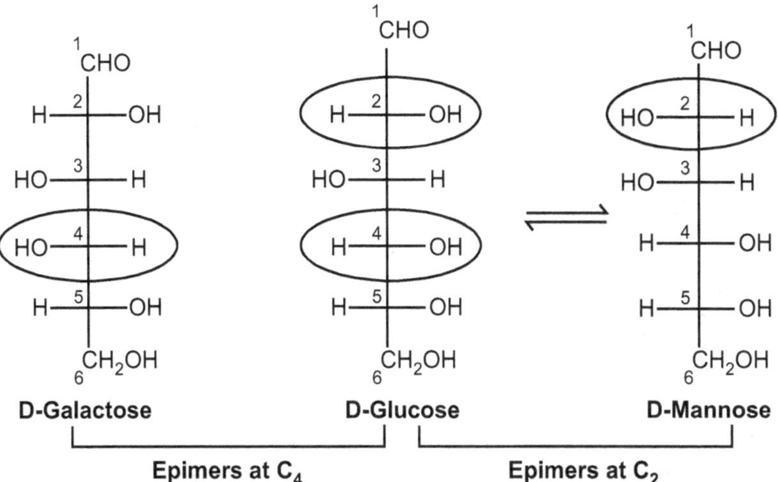

3.6.2 Anomers

Anomers and epimers are both diastereomers, but an epimer is a stereoisomer that differs in configuration at any single stereogenic center, while *'an anomer is actually an epimer that differs in configuration at the acetal/hemiacetal carbon'*.

'Anomerization is the process of conversion of one anomer to another anomer'. Two anomers are designated as alpha (α) or beta (β), according to the configurational relationship between the anomeric center and the anomeric reference atom. If the hydroxyl group on C_1 and the $-CH_2OH$ group on C_5 are on opposite sides of the six-membered ring, C_1 is said to be the α-anomer. If they are on the same side, C_1 is said to be the β-anomer.

Example 1: α-D-glucopyranose and β-D-glucopyranose.

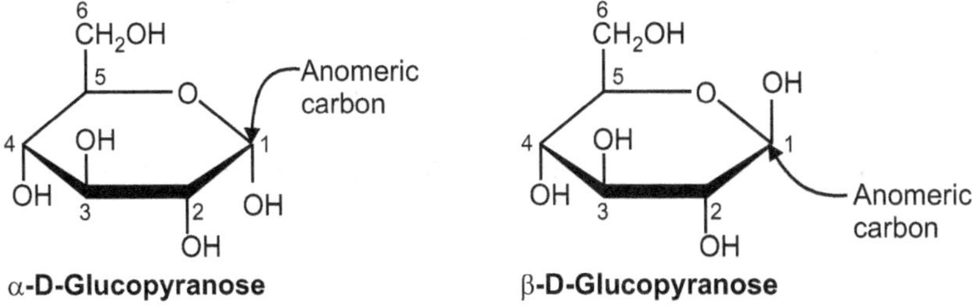

Example 2: α-D-fructofuranose and β-D-fructofuranose are anomers.

α-**D-Fructofuranose**

β-**D-Fructofuranose**

Note that the two stereoisomers (in the figure above) differ from each other in the configuration of C_1.

3.6.3 Mutarotation

Mutarotation is the change in the optical rotation because of the change in the equilibrium between two anomers, when the corresponding stereocenters interconvert.

Mutarotation was discovered by French chemist Dubrunfaut in 1844, when he noticed that the specific rotation of aqueous sugar solution changes with time.

Cyclic sugars show mutarotation as α and β anomeric forms interconvert. The optical rotation of the solution depends on the optical rotation of each anomer and their ratio in the solution.

In glucose, α and β isomers (anomers) differ in the orientation of –OH group at the C_1 hemiacetal carbon.

1. In the "alpha" (α) anomer of D-glucose, the OH group on C_1 is on the opposite side of the ring as the chain on C_5 with a specific rotation of +112° in water.

2. In the "beta" (β) anomer of D-glucose, the OH group on C_1 is on the same side of the ring as the C_5 substituent with a specific rotation of +19° (18.7 actually, but rounding upto 19).

Here is the interesting thing. When either anomer is dissolved in water, the value of the specific rotation changes over time, eventually reaching the same value of +52.5°.

*   The specific rotation of α-D-glucopyranose decreases from +112° to +52.5°.

*   The specific rotation of β-D-glucopyranose increases from +19° to +52.5°.

This behaviour is called **mutarotation** ('Muta' means 'change', so it literally means a change in rotation. The equilibrium mixture is actually about 64% of β-D-glucopyranose and about 36% of α-D-glucopyranose in water at 25°C with trace of open chain structure.

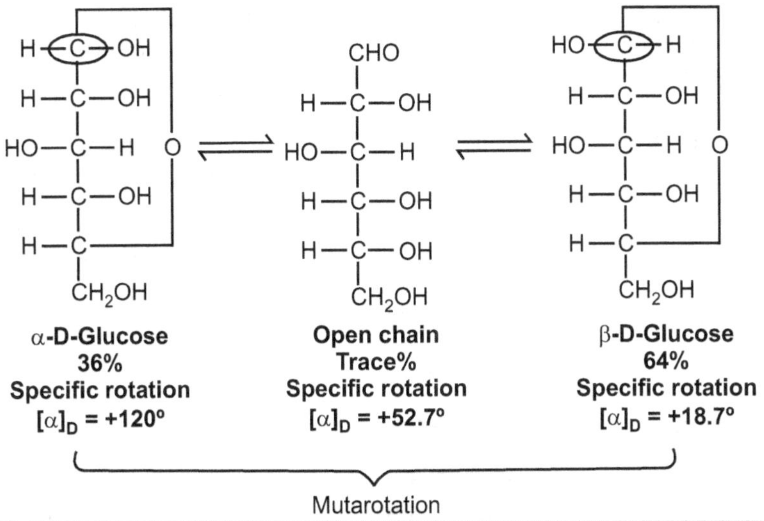

α-D-Glucose
36%
Specific rotation
$[\alpha]_D = +120°$

Open chain
Trace%
Specific rotation
$[\alpha]_D = +52.7°$

β-D-Glucose
64%
Specific rotation
$[\alpha]_D = +18.7°$

Mutarotation

3.7 Determination of Size of the Ring of Glucose by Methylation Method

The names furanose and pyranose have been coined to denote 5- and 6-membered rings in cyclic sugars. As glucose is 6-membered cyclic sugar, the two forms of glucose are appropriately identified by the names α-D-glucopyranose and β-D-glucopyranose. The presence of six membered oxide (pyranose) ring was determined by two methods: 1. Methylation method and 2. Periodic acid method.

Methylation/Haworth method: This is a classical method for determination of ring size in which ring sizes (furanose or pyranose) of D-Glucose have been determined using alkylation as a key step. It involves following steps;

1. **Methylation of anomeric hydroxyl (−OH) group:** The acid (dry HCl)-catalyzed methylation of glucose with methanol gives methyl-D-glucoside, due to displacement of the hemiacetal hydroxyl by methoxyl to form an acetal.

2. **Methylation of other four hydroxyl (−OH) groups:** The remaining four hydroxyl groups can be methylated in basic solution by dimethyl sulfate (or methyl iodide and silver oxide) to form pentamethyl glucose derivatives (i.e. methyl tetra-o-methyl-D-glucoside).

3. **Hydrolysis of pentamethyl glucose:** The pentamethyl glucose derivatives on hydrolysis with aqueous acid (dil. HCl) affects only the acetal linkage and leads to Tetra-o-methyl-D-glucose.

4. **Oxidation of Tetra-o-methyl-D-glucose (formation of lactone):** Tetra-o-methyl-D-glucose on oxidation with Br_2-water forms Tetra-o-methyl-glucono lactone.

5. **Oxidation of Tetra-o-methyl-glucono lactone:** Tetra-o-methyl- glucono lactone on oxidation with hot HNO_3 gives trimethoxypentanedioic acid (xylotrimethoxy glutaric acid).

$$D\text{-Glucose} \xrightarrow[\text{MeOH}]{\text{dry HCl}} \text{Methyl-D-glucoside} \xrightarrow[\text{NaOH}]{4Me_2SO_4} \text{Methyl tetra-o-methyl-D-glucoside}$$

$$\xrightarrow[\text{H}_2\text{O}]{\text{dil HCl}} \text{Tetra-o-Methyl-D-glucose} \xrightarrow[\text{H}_2\text{O}]{Br_2} \text{Tetra-o-methyl-glucono lactone}$$

$$\xrightarrow{HNO_3}$$

$$\begin{array}{c}
\text{COOH} \\
\text{H}—\!\!\!—\text{OCH}_3 \\
\text{H}_3\text{CO}—\!\!\!—\text{H} \\
\text{H}—\!\!\!—\text{OCH}_3 \\
\text{COOH}
\end{array}$$

Xylotrimethoxy glutaric acid

The formation of xylotrimethoxy glutaric acid due to the cleavage of C_5-C_6 bond of 1, 5-ring gives xylotrimethoxy glutaric acid (the groups which engage the ring are converted to the carboxylic group on oxidation). This indicates that the two carboxyl carbons must have been the ones originally involved in ring formation, and the oxide ring must be between C_1 and C_5. That is D-glucose has six-membered/pyranose/1 : 5 ring.

Reactions:

$$\begin{array}{c}
\text{H}—\text{C*}—\text{OH} \\
\text{H}—\text{C}—\text{OH} \\
\text{HO}—\text{C}—\text{H} \quad \text{O} \\
\text{H}—\text{C}—\text{OH} \\
\text{H}—\text{C*}— \\
\text{CH}_2\text{OH}
\end{array} \xrightarrow[\text{MeOH}]{\text{dry HCl}} \begin{array}{c}
\text{H}—\text{C*}—\text{OCH}_3 \\
\text{H}—\text{C}—\text{OH} \\
\text{HO}—\text{C}—\text{H} \quad \text{O} \\
\text{H}—\text{C}—\text{OH} \\
\text{H}—\text{C*}— \\
\text{CH}_2\text{OH}
\end{array} \xrightarrow[\text{NaOH}]{4Me_2SO_4} \begin{array}{c}
\text{H}—\text{C*}—\text{OCH}_3 \\
\text{H}—\text{C}—\text{OCH}_3 \\
\text{H}_3\text{CO}—\text{C}—\text{H} \quad \text{O} \\
\text{H}—\text{C}—\text{OCH}_3 \\
\text{H}—\text{C*}— \\
\text{CH}_2\text{OCH}_3
\end{array}$$

D-Glucose   Methyl-D-Glucoside   Methyl tetra-o-methyl-D-glucoside

$$\xrightarrow[\text{H}_2\text{O}]{\text{dil HCl}} \begin{array}{c}
\text{H}—\text{C*}—\text{OH} \\
\text{H}—\text{C}—\text{OCH}_3 \\
\text{H}_3\text{CO}—\text{C}—\text{H} \quad \text{O} \\
\text{H}—\text{C}—\text{OCH}_3 \\
\text{H}—\text{C*}— \\
\text{CH}_2\text{OCH}_3
\end{array} \xrightarrow[\text{H}_2\text{O}]{Br_2} \begin{array}{c}
\text{*C}=\text{O} \\
\text{H}—\text{C}—\text{OCH}_3 \\
\text{H}_3\text{CO}—\text{C}—\text{H} \quad \text{O} \\
\text{H}—\text{C}—\text{OCH}_3 \\
\text{H}—\text{C*}— \\
\text{CH}_2\text{OCH}_3
\end{array} \xrightarrow{HNO_3} \begin{array}{c}
\text{C*OOH} \\
\text{H}—\!\!\!—\text{OCH}_3 \\
\text{H}_3\text{CO}—\!\!\!—\text{H} \\
\text{H}—\!\!\!—\text{OCH}_3 \\
\text{C*OOH}
\end{array}$$

Tetra-o-methyl-D-glucose   Tetra-o-methyl-glucono lactone   Xylotrimethoxy glutaric acid

3.8 Representation of Ring Structure of Glucose

The ring structure of glucose can be represented in following two ways:

1.   Fischer projection formula: A Fischer projection represents a three dimensional structure in two dimensional form. In a Fischer projection, the longest chain is drawn vertically. The other bonds to a chiral carbon make a cross, with the carbon atom at the intersection of the horizontal and vertical lines. The two horizontal bonds are pointing toward the viewer. The two vertical bonds are directed away from the viewer.

In this form, place the most oxidizable group at the top. Place the CH_2OH group at the bottom of the Fischer projection which is not chiral.

2.   Haworth projection: A common way of representing the cyclic structure of monosaccharides is the Haworth projection, named after the English chemist Sir Walter N. Haworth (1937 Nobel Prize for chemistry). In a Haworth projection, a six-membered cyclic hemiacetal is represented as a planar hexagon, lying perpendicular to the plane of the paper. Groups bonded to the carbons of the ring then lie either above or below the plane of the ring. The new chiral center created in forming the cyclic structures is called an anomeric carbon. Haworth projections are most commonly written with the anomeric carbon to the right and the hemiacetal oxygen to the back. It involves following steps:

Step-I: The first step to convert a Fischer to a Haworth is to draw the wedges and number the carbons.

Step-II: Next, turn this molecule on its side by 90° clockwise and make a ring between the C_5-OH and the carbonyl carbon (C_1).

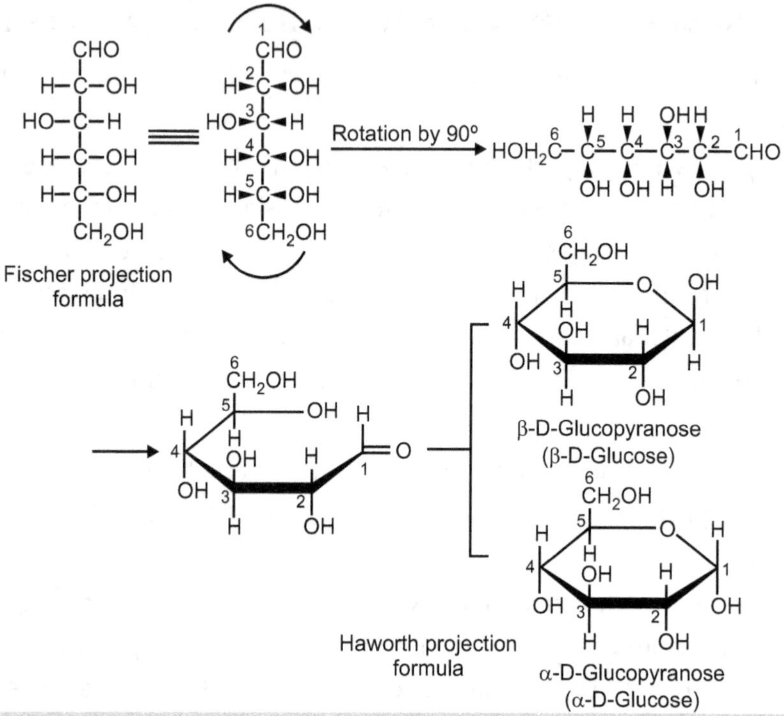

Fischer projection formula

Rotation by 90°

Haworth projection formula

β-D-Glucopyranose
(β-D-Glucose)

α-D-Glucopyranose
(α-D-Glucose)

3.9 Cyclic Structure of Fructose

Fructose is a 6-carbon polyhydroxyketone monosaccharide. The chemical formula of fructose is $C_6H_{12}O_6$ but the bonding of fructose is very different than that of glucose.

Like glucose, fructose also has a cyclic structure. In its cyclic form, it (generally) forms a five-membered ring which we call a furanose ring.

1. Since fructose contains a keto group, it forms an intramolecular hemiketal.

2. In the hemiketal formation, C_5 –OH of the fructose combines with C_2–keto group.

3. As a result, C_2 becomes chiral and thus has two possible arrangements of CH_2OH and OH group around it.

Two cyclic forms are possible because, in going from a straight chain to a ring, a new asymmetric carbon is introduced at carbon C_2 (the hemiketal carbon). Thus, D-fructose exists in two stereoisomeric forms, i.e., α-D-fructopyranose and β-D fructopyranose.

The stereochemistry of the cyclic forms of D-fructose is often represented by their Haworth projection shown below.

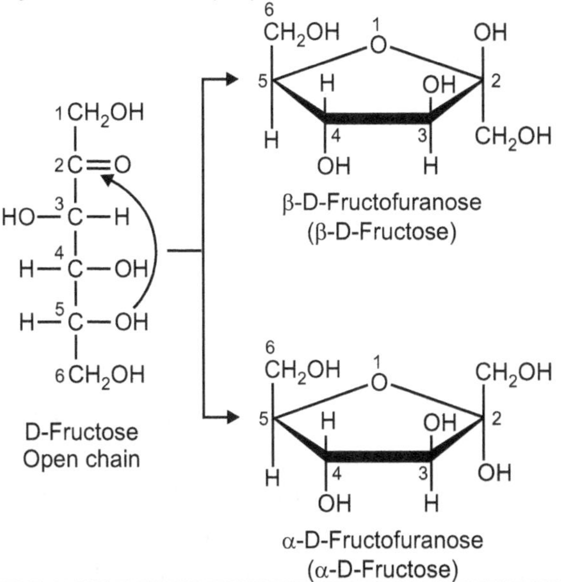

3.10 Structure of Disaccharides (Maltose, Lactose and Sucrose)

Types and Properties: The oligosaccharides which on hydrolysis form two same/different molecules of simple sugars are called disaccharides.

$$\text{Disaccharides} \xrightarrow[\text{H}_3\text{O}^+]{\text{Hydrolysis}} \text{2 Monosaccharides}$$

Same/different

In disaccharides, two sugar units are joined together by special type of ether linkages called 'acetal/glycosidic bonds' (C-O-C). Glycosidic bond is formed by condensation of an anomeric –OH group of one sugar unit with alcoholic/anomeric –OH of second sugar unit.

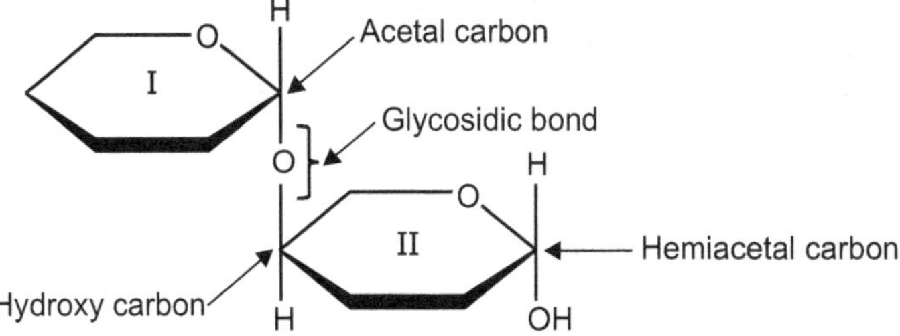

Common disaccharides: The most common type of disaccharides; sucrose, lactose, and maltose have 12 carbon atoms, with the general formula $C_{12}H_{22}O_{11}$. The differences in these disaccharides are due to atomic arrangements within the molecule.

Table 3.2

Disaccharides	Unit I	Unit II	Bond
Sucrose (*Table sugar*)	Glucose	Fructose	$\alpha(1 \rightarrow 2)\,\beta$ α, β-1,2 glycosidic bond
Lactose (*Milk sugar*)	Galactose	Glucose	$\beta(1 \rightarrow 4)$ β-1,4-glycosidic bond
Maltose (*Malt sugar*)	Glucose	Glucose	$\alpha(1 \rightarrow 4)$ α-1,4-glycosidic bond

1. Maltose (4'-O-(α-D-glucopyranosyl)-D-glucopyranose):

Maltose known as malt sugar is a disaccharide formed from two glucose molecules, which are joined together at carbons C_1 of one glucose and C_4 of other glucose called α-1,4-glycosidic bond.

Structure: It has the molecular formula $C_{12}H_{22}O_{11}$ and its systematic name is 4'-O-(α-D-glucopyranosyl)-D-glucopyranose. Maltose is reducing sugar, and hence gives positive test to Tollen's as well as Fehling's solution.

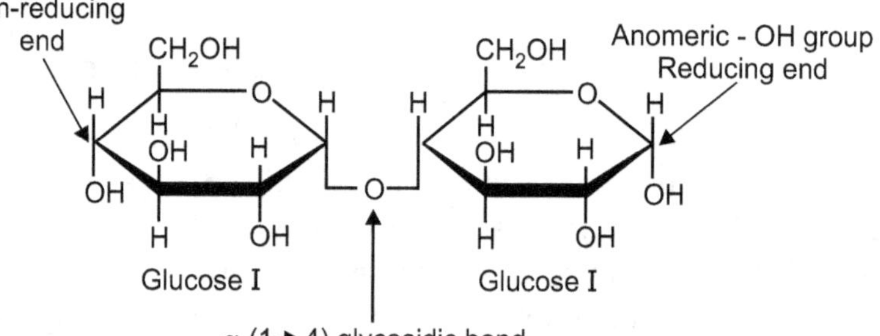

Hydrolysis: Maltose is produced when the enzyme amylase breaks down starch. When hydrolyzed, maltose yields two molecules of glucose.

$$(C_6H_{10}O_5)_n + H_2O \xrightarrow[\text{Partial hydrolysis}]{\beta\text{-amylase}} C_{12}H_{22}O_{11} \xrightarrow[\text{Hydrolysis}]{\text{Maltase}} C_6H_{12}O_6$$

Starch   Maltose   Glucose

Occurrence: Maltose is a white crystalline solid and does not occur in the average diet. It occurs in sprouting grains, malted cereals and malted milk. Among sweetening agent, this is found in corn syrup and corn sugar. In the body the most important source of maltose is as an intermediate product in the digestion of starch.

Uses:

1. It is used in confectionery and for preparation of culture medium for bacterial growth.
2. Maltose is formed in germinating cereal grains and is important in the production of alcohol by fermentation.

2. Lactose (4'-O-(β-D-galactopyranosyl)-D-glucopyranose):

Lactose is a disaccharide composed of galactose and glucose subunits. It is derived from the condensation of galactose and glucose, which form β-1 → 4 glycosidic linkage. The name comes from '*lac*' the Latin word for milk, plus the suffix -ose used to name sugars.

Structure: It has the molecular formula $C_{12}H_{22}O_{11}$ and its systematic name is β-D-galactopyranosyl-(1→4)-D-glucose. Lactose is a reducing sugar, and hence gives positive test to Tollen's as well as Fehling's solution.

Hydrolysis: Lactose on hydrolysis by dil. acid or enzyme lactase forms D-glucose and D-galactose.

$$C_{12}H_{22}O_{11} + H_2O \xrightarrow[\text{Hydrolysis}]{\text{Lactase}} C_6H_{12}O_6 + C_6H_{12}O_6$$

Lactose   Glucose   Galactose

Occurrence: Lactose makes up around 2–8% of milk (by weight).

The compound is a white, water-soluble, non-hygroscopic solid with a mildly sweet taste. Dairy products such as yogurt, cream and fresh cheeses have lactose contents similar to that of milk. Ripened cheeses contain little to no lactose, as bacteria convert most of it into lactic acid during the ripening process.

Uses:

1. Its mild flavor and easy handling properties have led to its use as a carrier and stabilizer of aromas and pharmaceutical products.
2. Lactose is added to tablet and capsule drug products as an ingredient because of its physical and functional properties, i.e., compressibility and cost effective use.
3. Lactose is not added directly to many foods, because its solubility is less than that of other sugar commonly used in food.
4. Lactose may be used to sweeten the stout beer; the resulting beer is usually called a milk stout or a cream stout.

3. Sucrose (α-D-glucopyranosyl-(1→2)-β-D-fructofuranoside):

It is a disaccharide, composed of two monosaccharide units: glucose and fructose which are connected with each other by α, β-1,2 glycosidic bond.

Structure: It has the molecular formula $C_{12}H_{22}O_{11}$ and is also called α-D-glucopyranosyl-(1→2)-β-D-fructofuranoside. In sucrose, the components glucose and fructose are linked via an ether bond between C_1 on the glucose subunit and C_2 on the fructose unit. The anomeric carbons of both glucose and fructose must be linked through an oxygen bridge in sucrose. Hence, sucrose is not a reducing sugar.

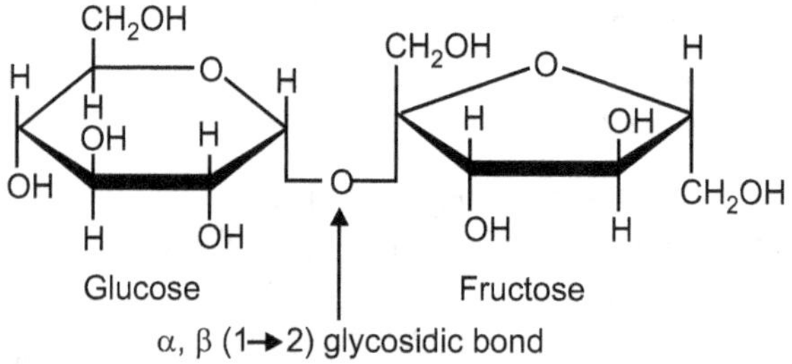

Hydrolysis: Sucrose on hydrolysis by dil. acid or enzyme invertase (sucrose) forms D-glucose and D-fructose. The optical rotataion changes from dextro to laevo, so sucrose is called as invert sugar.

$$C_{12}H_{22}O_{11} + H_2O \xrightarrow[\text{Hydrolysis}]{\text{Invertase}} C_6H_{12}O_6 \quad + \quad C_6H_{12}O_6$$

Sucrose   D-Glucose   D-fructose

$[\alpha]_D = +66.5°$ $[\alpha]_D = +52.7°$ $[\alpha]_D = -92.4°$

Occurrence: Sucrose is produced naturally in plants, from which table sugar is refined. It is manufactured from sugar canes or beet.

Uses:

1. High-fructose corn syrup (HFCS) is significantly cheaper as a sweetener for food and beverage than refined sucrose.
2. It is used as a sweetening agent in pharmaceuticals.
3. It is used as a sweetening agent in jams, jellies and confectionaries.

3.11 Polysaccharides (Cellulose and Starch)

Polysaccharides are relatively complex carbohydrates. They are made up of many monosaccharides joined together by glycosidic bonds.

Example: Cellulose and Starch.

1. Cellulose:

Cellulose is a linear homopolymer of glucose, in which several hundred to many thousands of glucose monomers are joined by β-1,4 linkages/bonds.

Structure: Cellulose is an organic compound with the molecular formula $(C_6H_{10}O_5)_n$.

Glucose — Non-reducing end   Glucose β (1→4) glycosidic bond   Glucose   Glucose — Reducing end

Hydrolysis: Cellulose on hydrolysis by strong acids (HCl and H_2SO_4) or enzyme cellulase forms D-glucose. Indeed, enzymatic hydrolysis of cellulose leads to cellobiose.

$$(C_6H_{10}O_5)_n + H_2O \xrightarrow[\text{Hydrolysis}]{\text{Cellulase}} n\ C_6H_{12}O_6$$

Cellulose   D-Glucose

Occurrence and Function: Cellulose is an important structural component of the primary cell wall of green plants, many forms of algae. The fibrous tissue in the cell walls of plants contains cellulose, consists of long chains of glucose units. The molecular weight of cellulose varies with the sources, but is usually high. Cotton cellulose appears to have about 3000 glucose units per molecule.

Uses:

1. For industrial use cellulose is mainly obtained from wood pulp and cotton.

2. Cellulose is converted into biofuels such as cellulosic ethanol which is used as a renewable fuel source.

3. Cellulose is used as fabric and also used to prepare acetate rayon and cellulose xanthate.

4. It is used to prepare gun cotton used as detonator.

2. Starch:

A second very widely distributed polysaccharide is starch. It is composed of two structurally different polysaccharides (both consist entirely of glucose units) one is amylose (linear structure) and the other is amylopectin (branched structure).

Structure: The chemical composition of starch varies with the source, but in any one starch there are two structurally different polysaccharides. Both consist entirely of glucose units, but one is a linear structure (amylose) and the other is a branched structure (amylopectin). Natural starches consist of about 10%-30% amylose and 70%-90% amylopectin.

(I) Amylose: The amylose is a linear homo glucopolymer consists of repeating 100-300 units of D-glucose held together by α-1,4-glycosidic bonds. It is also called as 'polymaltose'. It contains both reducing and non-reducing ends.

CH_2OH CH_2OH CH_2OH CH_2OH

Glucose   Glucose   Glucose   Glucose

Non-reducing end   α (1→ 4) glycosidic bond   Reducing end

(II) Amylopectin: In amylopectin, amylose chains are joined together by α-1,6 linkages. The branching occurs at about 25-30 glucose units. It contains nearly 1000 glucose units per amylopectin.

Glucose Glucose
CH_2OH CH_2OH
Amylose unit

Branching
β (1→ 6) glycosidic bond

CH_2OH CH_2OH H_2C CH_2OH
Amylose unit

Glucose Glucose Glucose Glucose

α (1→ 4) glycosidic bond

Hydrolysis: The hydrolysis does not proceed all the way to glucose because the β-1,6 glucosidic link at the branch point is not easily hydrolyzed. The complete hydrolysis of starch yields (in successive stages) glucose: starch → dextrins → maltose → glucose.

Whenever starch molecules undergo hydrolysis, it forms either monosaccharides, disaccharides or trisaccharides. The end products depend on the strength of enzymes.

The common enzymes e.g. β-Amylase, produces the disaccharide maltose while, γ-Amylase, produces glucose. Amylopectin on complete hydrolysis gives D-glucose.

Enzymes also catalyze hydrolysis of starch, but the enzyme amylase is specific for α-1,4 links and like acid-catalyzed hydrolysis, give a mixture of glucose, maltose, and polysaccharides (dextrin). The enzyme glucosidase can hydrolyze the β-1,6 glucosidic links at the branch points and, when used in conjunction with amylase, completes the hydrolysis of starch to glucose.

Occurrence: Starch is a potential source of glucose, which is stored in seeds, roots, and fibers of plants as a food reserve. Animals also store glucose in the form of starch like substances called glycogens.

Applications:
1. Starch is used in paper manufacture and in textile and food industries.
2. Fermentation of grain starches is an important source of ethanol.
3. Hydrolysis of starch catalyzed by hydrochloric acid results in a syrupy mixture of glucose, maltose, and higher-molecular-weight saccharides. This mixture is called dextrin and is marketed as corn syrup.

Note: Experimental evidence indicates that amylose is not a straight chain of glucose units but instead is coiled like a spring (helix), with six glucose monomers per turn. When coiled in this fashion, amylose has just enough room in its core to accommodate an iodine molecule. The characteristic blue-violet colour that appears when starch is treated with iodine is due to the formation of the amylose-iodine complex. This colour test is sensitive enough to detect even minute amounts of starch in solution.

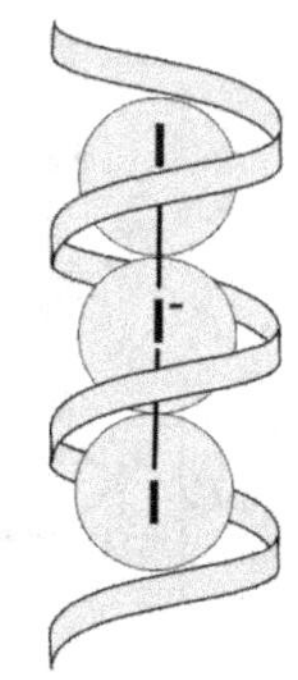

Structure of Starch-Iodine complex

Exercises

(A) Define the following terms with example:

1.	Mutarotation	2.	Anomers
3.	Epimers	4.	Anomeric carbon
5.	Reducing sugar	6.	Non-reducing sugars
7.	Monosaccharides	8.	Epimerization

(B) Short Answer Questions:

1. How do you explain the presence of all the six carbon atoms in glucose in a straight chain?
2. Under what conditions glucose is converted to gluconic and saccharic acid?
3. Which sugar is called invert sugar? Why is it called so?
4. How do you explain the presence of five-OH groups in glucose molecule?
5. How do you explain the presence of an aldehydic group in a glucose molecule?
6. Which monosaccharide units are present in starch, cellulose and which linkages link these units?
7. How will you distinguish 1° and 2° hydroxyl groups present in glucose? Explain with reactions.
8. Draw the cyclic structure for α and β-D-glucose. Identify the anomeric carbons.

9. Draw Haworth structures for each of the following disaccharides.

   (a) Sucrose, (b) Maltose, (c) Lactose, (d) Starch and (e) Cellulose.

10. What purposes do starch and cellulose serve in plants?

11. Define carbohydrates. Give scheme of its classification.

12. Write a note on mutarotation.

13. Explain difference between reducing and non-reducing sugars.

(C) Long Answer Questions:

1. Describe similarities and differences between starch and cellulose.

2. Write the reactions of D-glucose which cannot be explained by its open-chain structure. How can cyclic structure of glucose explain these reactions?

3. On the basis of various evidences assign complete structure of D-glucose.

4. What are polysaccharides? Explain the structure of cellulose and starch.

5. What are disaccharides? Explain the structure of maltose, lactose and sucrose.

6. Explain in detail reducing and non-reducing sugars.

7. Explain methylation method used for determination of ring size of glucose.

(D) Choose the correct alternative for each of the following and rewrite the sentence.

1. One molecule of sucrose on hydrolysis gives ....

   (a) 2 molecules of glucose

   (b) 1 molecule of glucose + 1 molecule of galactose

   (c) 1 molecule of glucose + 1 molecule of fructose

   (d) 2 molecules of fructose

2. Which of the following disaccharide is a non-reducing sugar?

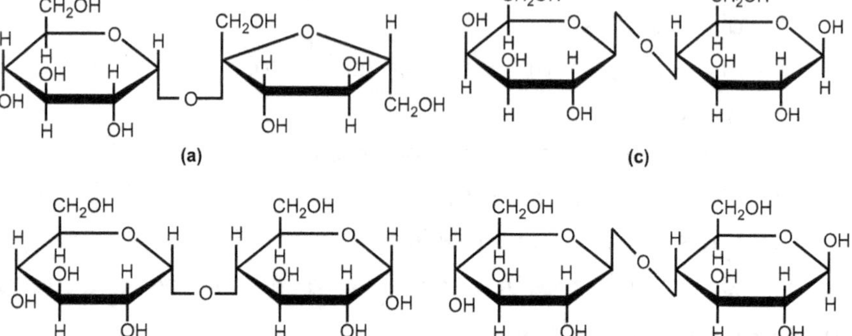

    (a)   (c)

    (b)   (d)

3. Which of the following statements is not true about glucose?
   (a) It is an aldohexose
   (b) On heating with HI it forms n-hexane
   (c) It is present in furanose form
   (d) It does not give phenylhydrazine test.

4. Three cyclic structures of monosaccharides are given below, which of these are anomers?

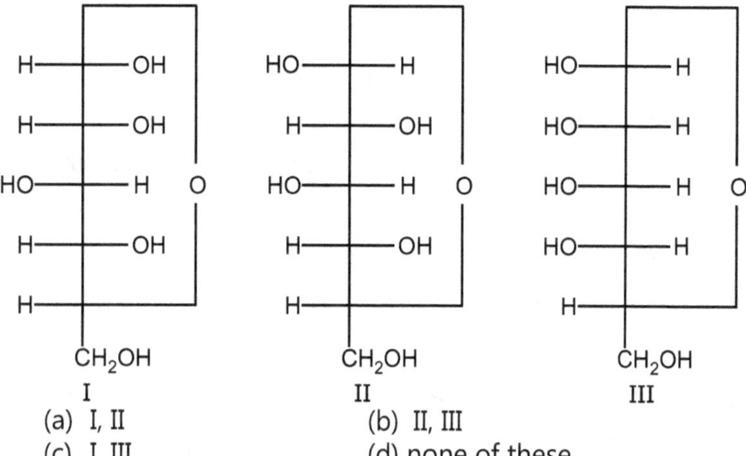

        I   II   III

   (a) I, II   (b) II, III
   (c) I, III   (d) none of these

5. Which of the following reactions of glucose can be explained only by its cyclic structure?
   (a) Glucose forms pentaacetate
   (b) Glucose reacts with hydroxylamine to form an oxime
   (c) Pentaacetate   of   glucose   does   not   react   with hydroxylamine.
   (d) Glucose is oxidised by nitric acid to gluconic acid

6. Optical rotations of some compounds along with their structures are given below, which of them have D configuration.

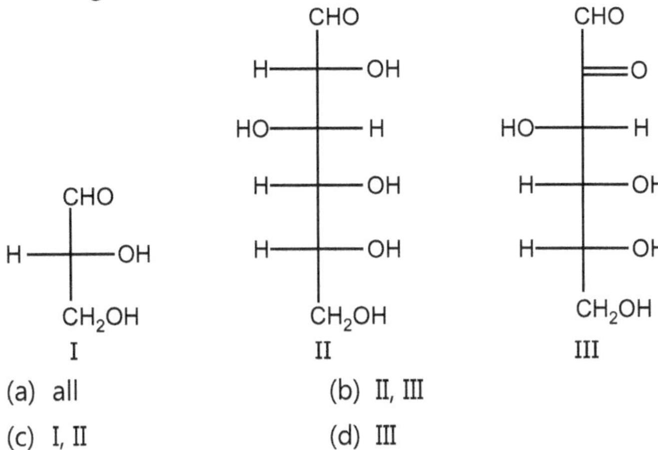

(a) all   (b) II, III

(c) I, II   (d) III

7. Structure of a disaccharide formed by glucose and fructose is given below. Identify anomeric carbon atoms in monosaccharide units.

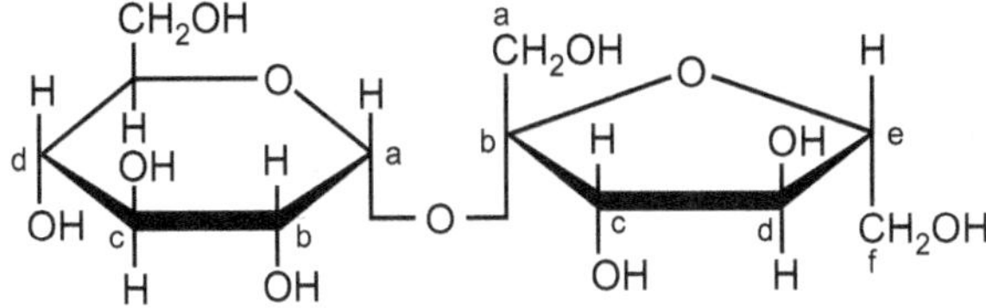

(a) 'a' carbon of glucose and 'a' carbon of fructose

(b) 'a' carbon of glucose and 'e' carbon of fructose

(c) 'a' carbon of glucose and 'b' carbon of fructose

(d) 'f' carbon of glucose and 'f' carbon of fructose

8. Carbohydrates are classified on the basis of their behavior on reducing or non-reducing sugar. Sucrose is a ......

(a) monosaccharide   (b) disaccharide

(c) reducing sugar   (d) non-reducing sugar

9. Which of the following carbohydrates are branched polymer of glucose?

(a) Amylose   (b) Amylopectin
(c) Cellulose   (d) Glycogen

10. Which of the following monosaccharides are present as five membered cyclic structure (furanose structure) ?
    (a) Ribose   (b) Glucose
    (c) Fructose   (c) Sucrose

11. Sucrose is a non-reducing sugar due to ......
    (a) 1-2 linkage   (b) 1-4 linkage
    (c) 1-5 linkage   (d) none of these

12. D – (+) glucose and maltose are ....
    (a) aldohexose and disaccharide
    (b) disaccharide and aldohexose
    (c) monosaccharide and polysaccharide
    (d) none of these

13. The sugar .... is invert sugar.
    (a) Ribose   (b) Glucose
    (c) Sucrose   (d) Galactose

14. Final product of methylation of glucose is........
    (a) Glucosazone   (b) Xylotriethoxy gluconic acid
    (c) Glucaric acid   (d) Xylotrimethoxy glucaric acid

15. Which one of the following is the reagent used to identify glucose?
    (a) neutral $FeCl_3$ (b) $CHCl_3$
    (c) Ammoniacal $AgNO_3$ (d) C_2H_5ONa

ANSWERS

1. (a)	2. (a)	3. (d)	4. (a)	5. (c)	6. (a)	7. (c)	8.(d)
9. (b)	10. (c)	11. (a)	12. (a)	13. (c)	14. (c)	15. (c)	

✱✱✱

Chapter **4**...

Carbonyl Compounds - Aldehydes and Ketones

Contents ...

4.1 Introduction

The compounds which contain a carbonyl group ($>C=O$) are collectively known as aldehydes and ketones. In aldehyde, the carbon atom of carbonyl group is bonded to one hydrogen atom and one organic group (R – CHO). While in ketone, it is bonded to two organic groups (RCOR). These are two important classes of carbonyl compounds which have many reactions in common, but have significant difference.

Many natural products contain these functional groups. For example, Benzaldehyde (Almonds), Cinnamaldehyde (Cinnamon), Testosterone (Sex hormone), Camphor (Camphor-Tree), Vanillin (Vanilla), Civetone (Civet-cats) etc.

The aldehydes and ketones formed two series of compounds having general formula $C_nH_{2n}O$. They have following structural formula,

$$R—\overset{\overset{\displaystyle O}{\|}}{C}—H \qquad R—\overset{\overset{\displaystyle O}{\|}}{C}—R'$$

Aldehyde   Ketone

(R = Alkyl or Aryl group)

The ketones in which two alkyl groups R and R' are same are called as simple ketones. When the two alkyl groups are different then these ketones are referred as mixed ketones.

4.1.1 Aldehydes

These are derivatives of hydrocarbons obtained by replacing two H-atoms of the end carbon atom of the aliphatic chain by an oxygen atom. They are classified as,

1. **Aliphatic aldehydes:** The aldehydes derived from aliphatic hydrocarbons are called aliphatic aldehydes.

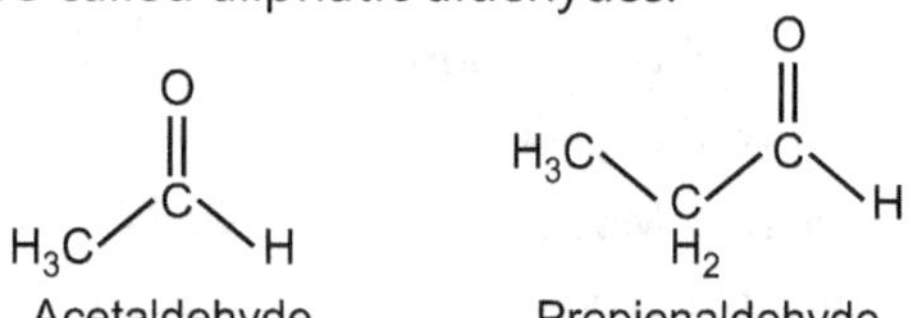

Acetaldehyde   Propionaldehyde

2. **Aromatic aldehydes:** Aldehydes derived from aromatic hydrocarbons are known as aromatic aldehydes.

Benzaldehyde   Phenyl acetaldehyde

4.1.2 Ketones

The ketones are derivatives of hydrocarbons and obtained by replacing two H-atoms of the central carbon atom by an oxygen atom. They are classified as follows.

1. **Aliphatic ketones:** The ketones which are derived from aliphatic hydrocarbons are known as aliphatic ketones.

Acetone   Ethyl methyl ketone

2.   Aromatic ketones: The ketones which are derived from aromatic hydrocarbons are known as aromatic ketones.

Benzophenone   Acetophenone

4.2 Nomenclature

(i) For Aldehydes:

Both common and IUPAC (International Union of Pure and Applied Chemistry) names are frequently used for aldehydes and ketones.

1.   Common system: The aldehydes are named after the acid produced by the oxidation of aldehyde. Thus, the aldehydes forming formic acid, acetic acid and propionic acid are named as formaldehyde, acetaldehyde and propionaldehyde respectively.

2.   IUPAC system: The aldehydes are named as the derivatives of corresponding hydrocarbons. Here are some simple IUPAC rules for naming aldehydes.

1.   The last letter 'e' of parent hydrocarbon is replaced by suffix *al.*

2.   The names of aldehydes are derived from those of the parent alkanes, defined by the longest continuous chain (LCC) of carbon atoms that contain the functional group.

3.   To indicate the position of a substituent on an aldehyde, the carbonyl carbon atom is always considered to be C_1 (i.e. chain numbering normally starts from the end nearest to the carbonyl group).

Table 4.1

Sr. No.	Formula	Common name	IUPAC name
1.	$H-\overset{\overset{O}{\|\|}}{C}-H$	Formaldehyde	Methanal
2.	$H_3C-\overset{\overset{O}{\|\|}}{C}-H$	Acetaldehyde	Ethanal

Sr. No.	Formula	Common name	IUPAC name
3.	$H_5C_2 - \overset{\overset{\displaystyle O}{\|\|}}{C} - H$	Propionaldehyde	Propanal
4.	$H_7C_3 - \overset{\overset{\displaystyle O}{\|\|}}{C} - H$	Butyraldehyde	Butanal
5.	$H_9C_4 - \overset{\overset{\displaystyle O}{\|\|}}{C} - H$	Valeraldehyde	Pentanal

Other examples:

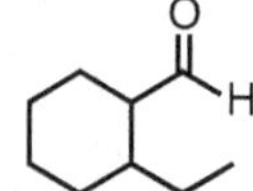

3-Bromobenzaldehyde   4-Bromopentanal   3-Bromo-4-methyl-pentanal

3-3-Dimethyl-butanal   3-Hydroxy-2, 4-dimethyl-pentanal   3-Chloro-3-phenyl-propenal

2-Ethyl-cyclohexanecarbaldehyde   3-Hydroxy-cyclopentanecarbaldehyde

(ii) For Ketones:

1.   Common system: Except in a few cases, radico functional names are used. The names of radicals attached to the carbonyl function are followed by ketone.

2.   IUPAC system: They are named as the derivatives of alkanes by replacing final letter 'e' of parent alkane by suffix **one**.

Table 4.2

Sr. No.	Formula	Common name	IUPAC name
1.	$H_3C - \overset{\overset{\displaystyle O}{\|\|}}{C} - CH_3$	Acetone (Dimethyl ketone)	2-Propanone

Sr. No.	Formula	Common name	IUPAC name
2.	$\overset{\displaystyle O}{\overset{\displaystyle \|}{H_3C - C - C_2H_5}}$	Ethyl methyl ketone	2-Butanone
3.	$\overset{\displaystyle O}{\overset{\displaystyle \|}{H_5C_2C - C_2H_5}}$	Diethyl ketone	2-Pentanone
4.	$\overset{\displaystyle O}{\overset{\displaystyle \|}{H_7C_3 - C - C_2H_5}}$	Ethyl propyl ketone	3-Hexanone
5.	$\overset{\displaystyle O}{\overset{\displaystyle \|}{H_7C_3 - C - C_3H_7}}$	Dipropyl ketone	4-Heptanone

2-bromo-4, 4-dimethylcyclohexanone

2-bromo-3-hydroxy-1-phenylbutanone

2,3-hexanedione

3-cyclobutyl-3-oxopropanol

3-methyl-2-butanone

2-cyclohexenone

4.3 Structure of a Carbonyl Group

The carbon oxygen bond in carbonyl group is composed of a sigma (σ) as well as pi (π) bonds, similar to alkene but instead containing two atoms of quite different electronegativities. The σ-bond is formed by overlap of sp^2-hybridized orbital of carbon with sp^2-hybridized orbital of oxygen atom. The π-bond is formed by side to sidewise (lateral) overlap of non-hybridized p_z-orbital of oxygen atom as shown in Fig. 4.1.

The carbon is sp^2 hybridized and three bonds sustain an angle of $120°$. Thus, all the bonds to the C-atom lie in the same plane and carbonyl bond ($>C=O$) distance is 1.24 A°. Since, oxygen is more electronegative than the C-atom, the shared pair electrons between them lie closer to the oxygen atom. The >C=O is thus polarized and because of this C-atom develops positive charge and it is attacked by many ionic reagents. The oxygen atom contains two pairs of non-bonding electrons and the electronic structure for carbonyl group is responsible for most of the properties of aldehydes and ketones.

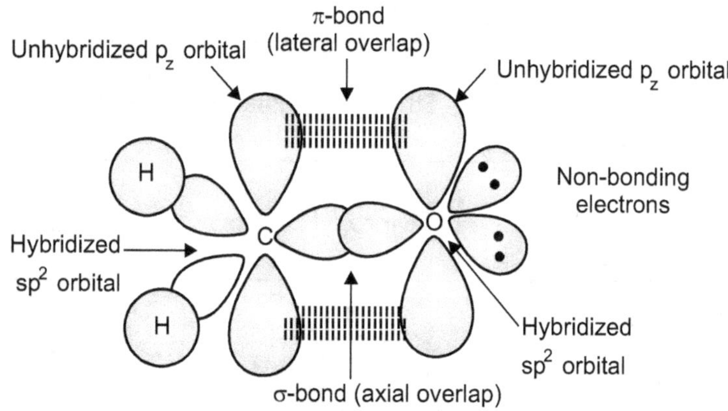

Fig. 4.1: Molecular hybrid picture of carbonyl group

4.4 Reactivity of Carbonyl Group

In carbonyl group, oxygen atom is connected to the carbon atom by double bond. The oxygen atom is more electronegative than carbon. Hence it attracts electrons towards itself, so electron concentration increases at oxygen atom to make it electronegative and carbon atom becomes electron deficient or electropositive. Hence, nucleophilic attack takes place at carbon atom more easily.

$$
\underset{\text{Polar structure}}{\overset{-\delta}{O} \quad \underset{\overset{+\delta}{C}}{\underset{R}{\overset{R}{\diagdown}}}} + Nu^- \longrightarrow \underset{\substack{\text{Unstable} \\ \text{tetrahedral intermediate}}}{\overset{O^-}{\underset{R}{\underset{R}{\overset{\mid}{C}}}}{-}Nu} \longrightarrow \underset{\text{Addition product}}{\overset{OH}{\underset{R}{\underset{R}{\overset{\mid}{C}}}}{-}Nu}
$$

4.4.1 Relative Reactivity

The reactivity of carbonyl group depends upon the nature of alkyl group (R). Due to the electron donating nature of alkyl group, the alkyl group donates electron to the carbonyl group. Due to this electron density increases and carbon becomes more electron rich. Hence, the ability of oxygen atom to take proton decreases and nucleophilic attack is retarded.

Formaldehyde   Acetaldehyde   Acetone

A second reason for less reactivity of carbonyl group is steric hindrance, the larger alkyl group is sterically hindered. Due to the larger size of alkyl group, nucleophilic attack is hindered.

The reactivity of some carbonyl compounds is given below.

Reactivity increases

Aromatic carbonyl compounds are less reactive than aliphatic carbonyl compound due to steric hindrance. The bigger phenyl group inhibits nucleophilic attack.

Acidic nature of α-hydrogen: The carbon atom adjacent to the carbonyl group is called as α-carbon atom and the hydrogen attached to α-carbon is called as α-hydrogen atom.

The polar nature of carbonyl group induces a partial positive charge on α-carbon atom, so α-carbon atom pulls the electron pair between α-carbon and α-hydrogen towards itself. This α-hydrogen atom is acidic, which may be removed to form carbanion, and are stabilized by resonance.

Labile   Aldehyde   Carbanion   Enolate ion

4.4.2 General Mechanism of Nucleophilic Addition to Carbonyl Group

The nucleophilic addition is observed in aldehydes and ketones. It takes place in two steps:

Step-I: Nucleophile attacks electrophilic carbon of carbonyl group to give an alkoxide ion.

Step-II: Protonation of alkoxide ion give final addition product.

4.5 Mechanism and Applications of Reactions

Aldehydes and ketones mainly give various types of nucleophilic addition reactions. Some of the reactions are given below.

(A) Aldol condensation (Base catalyzed)

Reaction: In the presence of base, aldehydes or ketones containing *α-hydrogen atom* undergo condensation to produce β-hydroxy aldehyde (aldol) or β-hydroxy ketone (ketol). This reaction is known as *aldol condensation*.

$$CH_3CHO + CH_3CHO \xrightarrow{\overset{\ominus}{OH}} CH_3-CH-\overset{\beta}{CH}-CH_2-CHO$$

Acetaldehyde

β-hydroxy aldehyde
(Aldol)

$$H_3C-\overset{O}{\overset{\|}{C}}-CH_3 + H_3C-\overset{O}{\overset{\|}{C}}-CH_3 \xrightarrow{\overset{\ominus}{OH}} H_3C-\overset{OH}{\overset{|}{\underset{|}{C}}}-CH_2-\overset{O}{\overset{\|}{C}}-CH_3$$

Acetone

4-methyl, 4-hydroxy, 2-pentanone
(Ketol)

Mechanism: Base catalyzed mechanism of aldol condensation involves following steps:

Step-I: Formation of carbanion: The base abstracts the α-hydrogen of ketone/aldehyde to form carbanion called enolate anion.

$$H-CH_2CHO + \overset{\ominus}{OH} \xrightarrow{-H_2O} \overset{\ominus}{CH_2}CHO$$

Acetaldehyde

Carbanion
(Nucleophile)

Step-II: Attack of carbanion on carbonyl group: Enolate anion attacks the carbonyl group of second aldehyde/ketone molecule to give anion I.

$$H_3C-\overset{O}{\underset{}{C}}-H \; + \; \overset{\ominus}{C}H_2CHO \longrightarrow H_3C-\overset{\overset{\ominus}{O}}{\underset{}{C}H}-CH_2-CHO$$

Acetaldehyde	Carbanion (Nucleophile)	Anion-I (Transition state)

Step-III: Formation of β-hydroxy aldehyde/ketone: Anion-I on protonation gives the β-hydroxy aldehyde.

$$H_3C-\overset{\overset{\ominus}{O}}{\underset{}{C}H}-CH_2-CHO \xrightarrow{\overset{\oplus}{H}} H_3C-\overset{OH}{\underset{}{C}H}-CH_2-CHO$$

Anion-I
(Transition state)   β-hydroxy aldehyde

Step-IV: Dehydration of β-hydroxy aldehyde: The dehydration of aldol in the presence of acid gives α, β-unsaturated aldehyde.

$$H_3C-\overset{OH}{\underset{}{C}H}-\overset{H}{\underset{}{C}H}-CHO \xrightarrow{\overset{\oplus}{H}} H_3C-CH=CH-CHO$$

Aldol   Butenal
                                  (Crotonaldehyde)

Applications:

(i) Synthesis of α, β-unsaturated aldehyde/ketone:

$$CH_3CHO + CH_3CHO \xrightarrow[\overset{\oplus}{H}]{\overset{\ominus}{O}H} H_3C-CH=CH-CHO$$

Acetaldehyde   Butenal
                                  (Crotonaldehyde)

$$H_3C-\overset{O}{\underset{}{C}}-CH_3 + H_3C-\overset{O}{\underset{}{C}}-CH_3 \xrightarrow[\overset{\oplus}{H}]{\overset{\ominus}{O}H} H_3C-\underset{\underset{CH_3}{|}}{C}=CH-\overset{O}{\underset{}{C}}-CH_3$$

α, β - unsaturated ketone

(ii) Synthesis of unsaturated acid:

$$C_6H_5CHO + CH_3CHO \xrightarrow[\overset{\oplus}{H}]{\overset{\ominus}{O}H} H_5C_6-CH=CH-CHO$$

Benzaldehyde Acetaldehyde   Cinnamaldehyde

$$\Big\downarrow (O)$$

$$H_5C_6-CH=CH-COOH$$

Cinnamic acid

$$CH_3CHO + CH_3CHO \xrightarrow{\overset{\ominus}{OH} / \overset{\oplus}{H}} H_3C-CH=CH-CHO$$

Acetaldehyde   Crotonaldehyde

$$\downarrow (O)$$

$$H_3C-CH=CH-COOH$$

Crotonic acid

(B) Perkin reaction:

Reaction: In the presence of base, aromatic aldehydes on condensation with anhydride of an aliphatic acid containing at least one *α-hydrogen* atom form α, β-unsaturated acid and the reaction is called as Perkin reaction.

It is carried out in the presence of sodium or potassium salt of an acid corresponding to anhydride.

$$C_6H_5CHO + (CH_3CO)_2O \xrightarrow[453 \text{ K}]{CH_3COONa} C_6H_5-CH=CH-COOH$$

Benzaldehyde   Acetic anhydride   Cinnamic acid

Mechanism: The base catalyzed mechanism of Perkin reaction involves following steps:

Step-I: Formation of carbanion: The base abstracts proton from acetic anhydride to form carbanion-I (acts as nucleophile).

$$CH_3\overset{\ominus}{COO} + H-CH_2-CO \cdots O \cdots H_3C-CO \longrightarrow H_2\overset{\ominus}{C}-CO \cdots O \cdots H_3C-CO + CH_3COOH$$

Acetic anhydride   Carbanion-I

Step-II: Formation of anion: Carbanion (nucleophile) attacks on the carbonyl carbon of benzaldehyde to form anion-II.

$$C_6H_5-CHO + H_2\overset{\ominus}{C}-CO \cdots O \cdots H_3C-CO \longrightarrow C_6H_5-\overset{\overset{\ominus}{O}}{CH}-CH_2-CO \cdots O \cdots H_3C-CO$$

Benzaldehyde   Carbanion-I   Anion-II

Step-III: Formation of product: Protonation of anion-II by acid formed in the step-II gives aldol type addition product.

Anion-II

Step-IV: The dehydration to give α, β-unsaturated anhydride.

Step-V: Hydrolysis of α, β-unsaturated anhydride to give α, β-unsaturated acid.

Cinnamic acid

Applications:

(i) Synthesis of substituted aromatic unsaturated acid:

Substituted unsaturated acid

(ii) Synthesis of Coumarin (Heterocyclic compound):

Salicylaldehyde   Coumarin

(C) Cannizzaro's reaction:

Reaction: In the presence of strong base, aldehydes lacking α-hydrogen atom undergo self oxidation and reduction to produce alcohol and acid. This is known as Cannizzaro's reaction.

This reaction is specially shown by aromatic aldehydes but by changing reaction condition other aldehydes can also show this reaction.

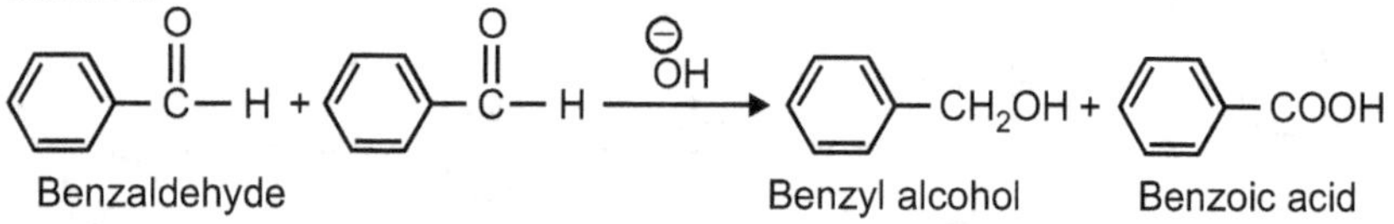

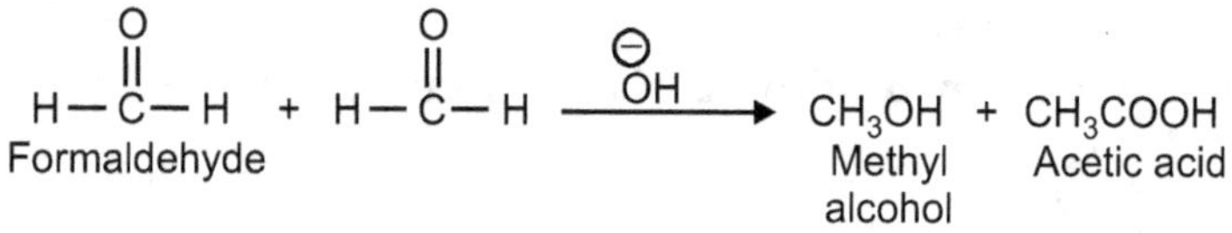

Benzaldehyde + Benzaldehyde $\xrightarrow{\overset{\ominus}{O}H}$ Benzyl alcohol (CH_2OH) + Benzoic acid (COOH)

Formaldehyde ($H-C(=O)-H$) + $H-C(=O)-H$ $\xrightarrow{\overset{\ominus}{O}H}$ CH_3OH (Methyl alcohol) + CH_3COOH (Acetic acid)

Mechanism: The mechanism of reaction is given below.

Step-I: Addition of hydroxide ion to the carbonyl group to give oxide ion.

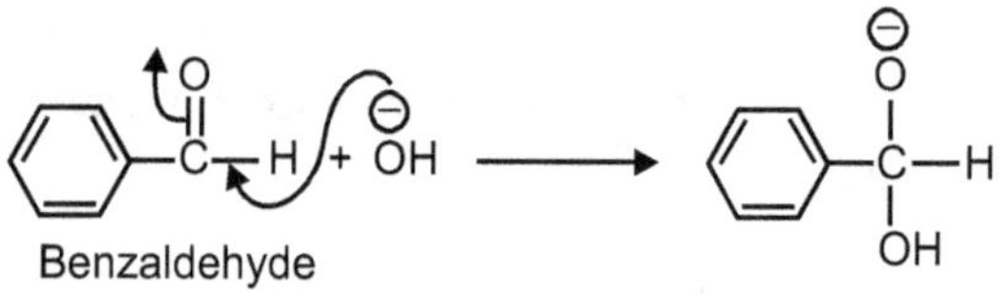

Benzaldehyde $+ \overset{\ominus}{O}H \longrightarrow$ [oxide ion intermediate]

Step-II: Transfer of intermolecular hydride ion.

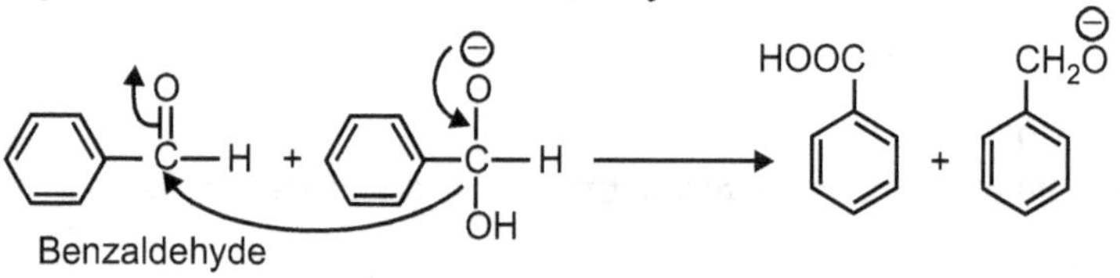

Benzaldehyde $+$ [intermediate] $\longrightarrow$ HOOC [ring] $+$ $CH_2\overset{\ominus}{O}$ [ring]

Step-III: Acid-base reaction gives salt of carboxylic acid and benzyl alcohol.

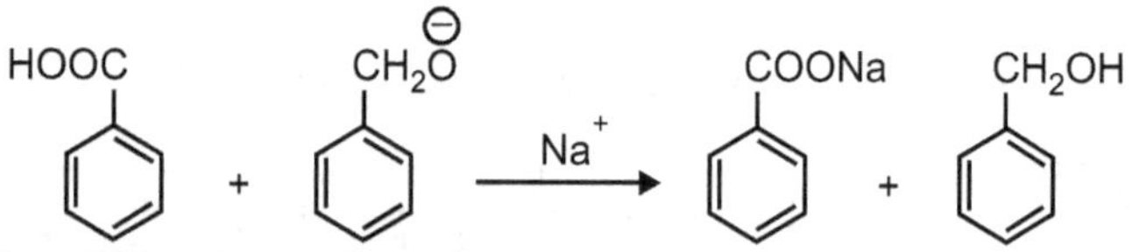

HOOC [ring] $+$ $CH_2\overset{\ominus}{O}$ [ring] $\xrightarrow{Na^+}$ COONa [ring] $+$ CH_2OH [ring]

Crossed or Mixed Cannizzaro's reaction:

Two different aldehydes lacking α-hydrogen atom in the presence of strong base like NaOH or KOH gives acid and alcohol and the reaction is called as Crossed Cannizzaro's reaction.

OHC [ring] (Benzaldehyde) $+$ $H-C(=O)-H$ (Formaldehyde) $\xrightarrow{NaOH}$ CH_2OH [ring] (Benzyl alcohol) $+$ HCOOH (Formic acid)

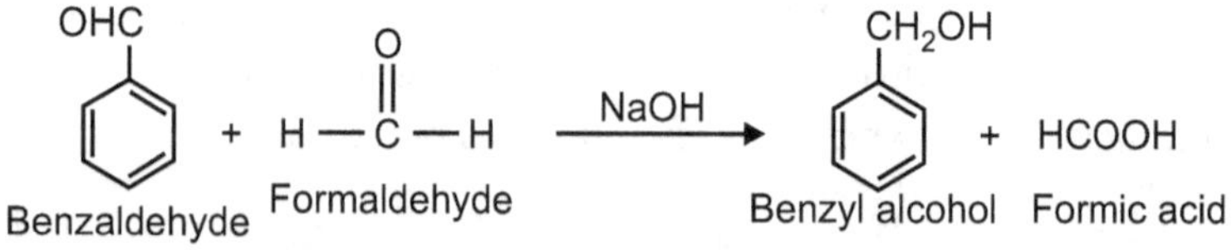

Applications:

(i) Synthesis of neopentyl glycol: Neopentyl glycol which is being used in polyesters, varnish coatings, synthetic lubricants and plasticizers.

Isobutyraldehyde　　　　Formaldehyde　　　　　Neopentyl glycol

(ii) Synthesis of substituted aromatic acid:

Phthalaldehyde　　　　　(o-hydroxymethyl) benzoic acid

(D) Knoevenagel condensation:

Reaction: When benzaldehyde is condensed with diethyl malonate in the presence of pyridine followed by hydrolysis and decarboxylation gives cinnamic acid and the reaction is called as Knoevenagel condensation.

Benzaldehyde　　　　　　Diethyl malonate　　　　　　　　Cinnamic acid

Mechanism: It is base catalyzed reaction and mechanism takes place as:

Step-I: Abstraction of proton from active methylene group by base.

Base　　　　Diethyl malonate　　　　　Carbanion (Nucleophile)

Step-II: Attack of nucleophile to the electron deficient carbonyl carbon of benzaldehyde.

Anion

Step-III: Protonation to give β-hydroxy diester.

Step-IV: Dehydration to give α, β-unsaturated diester.

Step-V: Hydrolysis of α, β-unsaturated diester to give α, β-unsaturated dicarboxylic acid.

Step-VI: Decarboxylation of α, β-unsaturated dicarboxylic acid to give α, β-unsaturated acid.

Cinnamic acid

Applications:

(i) Synthesis of substituted aromatic α, β-unsaturated acid:

Diethyl malonate

(ii) Synthesis of heterocyclic compound:

Diethyl malonate

3-carboethoxy coumarin-2-one

Salicylaldehyde

(E) Reformatsky reaction:

Reaction: Condensation reaction of carbonyl compound with alpha halo ester in the presence of zinc metal is known as Reformatsky reaction. The solvent most often used in this reaction is benzene, ether or benzene-ether mixture. The product of the reaction after acidification is β-hydroxy ester.

$$R_2C=O + X-CH_2-\overset{O}{\overset{\|}{C}}-OC_2H_5 \xrightarrow[\text{(ii) HOH/H}^+]{\text{(i) Zn/benzene}} R-\overset{OH}{\underset{R'}{\overset{|}{\underset{|}{C}}}}-CH_2-\overset{O}{\overset{\|}{C}}-OC_2H_5$$

Aldehyde or Ketone   α-haloester   β-haloester

Mechanism: It takes through following steps:

Step-I: Zinc is two electron donor and like to be oxidized from Zn(0) to Zn(II). Zinc reacts with α-halo ester to form zinc enolate.

$$Br-CH_2-\overset{O}{\overset{\|}{C}}-OC_2H_5 \longrightarrow \overset{\ominus}{Br}\ \ H_2C=\overset{O-\overset{+}{Zn}}{\overset{|}{C}}-OC_2H_5 \longrightarrow H_2C=\overset{O-Zn-Br}{\overset{|}{C}}-OC_2H_5$$

Zinc enolate

Step-II: Zinc enolate reacts cleanly with aldehydes and ketones to give zinc alkoxide which on hydrolysis gives β-hydroxy ester.

$$H_2C=\overset{O-Zn-Br}{\overset{|}{C}}-OC_2H_5 + RCHO \longrightarrow$$

Zinc enolate

$$R-\overset{OH}{\overset{|}{HC}}-CH_2-\overset{O}{\overset{\|}{C}}-OC_2H_5 \xleftarrow{\text{H}_2\text{O/H}^+}$$

Applications:

(i) Synthesis of α, β-unsaturated ester:

$$C_6H_5CHO + Br-\overset{CH_3}{\overset{|}{HC}}-\overset{O}{\overset{\|}{C}}-OC_2H_5 \xrightarrow[\text{(ii) HOH/H}^+/\Delta]{\text{(i) Zn/C}_6\text{H}_6/\Delta} H_3C-CH=\underset{CH_3}{\overset{|}{C}}-COOC_2H_5$$

α, β-unsaturated ester

(ii) Synthesis of β-keto ester:

$$CH_3C\equiv N + Br-H_2C-\overset{O}{\overset{\|}{C}}-OC_2H_5 \xrightarrow[\text{(ii) HOH/H}^+/\Delta]{\text{(i) Zn/Ether/}\Delta} H_3C-\overset{O}{\overset{\|}{C}}-CH_2-COOC_2H_5$$

Ketoester

Exercises

(A) Define with an example:

1. Aldol condensation
2. Cannizzaro's reaction
3. Reformatsky reaction

(B) Short Answer Questions (Write a note on):

(a) Aldol condensation
(b) Perkin reaction
(c) Cannizzaro's reaction
(d) Knoevenagel condensation
(e) Reformatsky reaction
(f) Structure of carbonyl group
(g) Reactivity of carbonyl group.

(C) Long Answer Questions:

1. Explain Perkin reaction giving its mechanism and application.
2. Explain the structure and reactivity of carbonyl group.
3. Explain Aldol condensation with mechanism and its uses.
4. Explain Cannizzaro's reaction with mechanism and its uses.
5. Explain Clemmensen reduction with mechanism and its uses.
6. Explain Reformatsky reaction with mechanism and its applications.
7. Explain Knoevenagel condensation with mechanism.

(D) Choose the Correct Alternative for each of the following and rewrite the sentence:

1. The hybridization of carbon atom in carbonyl group in aldehydes and ketones is respectively ....
   (a) sp^3 and sp^2
   (b) sp^3 and sp
   (c) sp^2 and sp^2
   (d) sp^2 and sp
2. Aldehydes are isomeric to ....
   (a) ketones
   (b) unsaturated alcohols
   (c) epoxides
   (d) all of these
3. Carbonyl group of aldehydes and ketones readily undergoes ....
   (a) electrophilic substitutions
   (b) nucleophilic substitutions
   (c) nucleophilic additions
   (d) nucleophilic substitutions

4. Reformatsky reaction is the reaction between a carbonyl compound, zinc and ....
    (a) unsaturated acid   (b) unsaturated ester
    (c) α-haloester   (d) β-haloester

5. The key step involved in Cannizzaro's reaction is ......
    (a) hydride ion transfer   (b) proton transfer
    (c) hydroxide ion transfer (d) carbocation transfer

6. The reaction intermediate involved in Perkin and Knoevenagel reaction is ....
    (a) carbocation   (b) carbanion
    (c) carbene   (d) carbonium ion

7. The carbonyl carbon of aldehydes and ketones bear ...... bonds.
    (a) one σ and two π (b) one π and two σ
    (c) one π and three σ (d) one σ and three π

8. Aldol condensation is carried out in the presence of ....
    (a) mild and dilute alkali
    (b) dilute acid
    (c) conc. alkali
    (d) conc. acid

9. ...... of the following undergoes Cannizzaro's reaction.
    (a) $CH_3.CHO$ (b) $C_6H_5.CH_2.CHO$
    (c) $HCHO$ (d) $CH_3.CO.CH_3$

10. Aldol condensation is mainly given by ....
    (a) aldehydes   (b) esters
    (c) α-haloesters   (d) carboxylic acids

11. In the presence of strong base, aldehyde lacking α-hydrogen atom undergo self oxidation and reduction to produce alcohol and acid and the reaction is known as ...... reaction.
    (a) Aldol   (b) Reformatsky
    (c) Perkin   (d) Cannizzaro

12. The product in Knoevenagel condensation is ....
    (a) alkene   (b) unsaturated acid
    (c) unsaturated aldehyde (d) unsaturated ketone

13. Which of the following does not have alpha hydrogen?
    (a) formaldehyde   (b) acetaldehyde
    (c) phenyl acetaldehyde   (d) acetophenone

14. The product of Perkin reaction is ....
    (a) cinnamic acid   (b) benzoic acid
    (c) salicylic acid   (d) β-hydroxy ester

15. The condensation reaction of carbonyl compound with alpha halo ester in the presence of zinc metal is known as ...... reaction.
    (a) Perkin   (b) Reformatsky
    (c) Knoevenagel   (d) Cannizzaro

ANSWERS

1. (c)	2. (d)	3. (c)	4. (c)	5. (a)	6. (b)	7. (c)	8. (a)
9. (c)	10. (a)	11. (d)	12. (b)	13. (a)	14. (a)	15. (b)	

***

Chapter 5...

Stereochemistry

Contents ...

5.1 Introduction

Nature is inherently chiral because the building blocks of life (amino acids, nucleotides, and sugars) are chiral and appear in nature in enantiomerically pure forms. Hence, any substances created by humankind to interact with or modify nature are interacting with a chiral environment. This is an important issue for bioorganic chemists and a practical issue for pharmaceutical chemists. The goal is for the student to gain a fundamental understanding of basic principles of stereochemistry and the associated terminology.

Stereochemistry is defined as a branch of chemistry which deals with study of three dimensional geometry of molecule and it's effect on physical and chemical properties of molecules. Simply it involves study of stereoisomers. It is the study of the static and dynamic aspects of the three-dimensional shapes of molecules. The geometry of molecule depends on various factors like bond angle, bond length, dihedral angle and orientation of atoms/groups in space.

5.2 Concept of Isomerism

In the study of organic compounds, there are many compounds having same molecular formula but different physical or chemical properties, such compounds are called *isomers* and the phenomenon is called *isomerism*. Since isomers have same molecular formula, the difference in their physical properties must be due to different arrangement of atoms within the molecule.

5.3 Types of Isomers and Isomerism

There are two types of isomerism:

(a) Structural isomerism.

(b) Stereoisomerism.

(a) Structural Isomerism: *The phenomenon in which there is difference in the arrangement of atoms within the molecule without any reference to space, is termed as structural isomerism.* In other words, they have same molecular formulas but have different structural formulas.

(b) Stereoisomerism: *The phenomenon which is caused by different arrangements of atoms or groups in spaces, is called as stereoisomerism.* In other words, these stereoisomers have same molecular as well as structural formulas but differ in the spatial arrangement of atoms or groups in the space. This type of isomerism includes:

(i) Configurational isomerism (Optical and Geometrical): The stereoisomers that interconvert through the breaking and reforming of covalent bonds, is a high-energy process. **Example:**

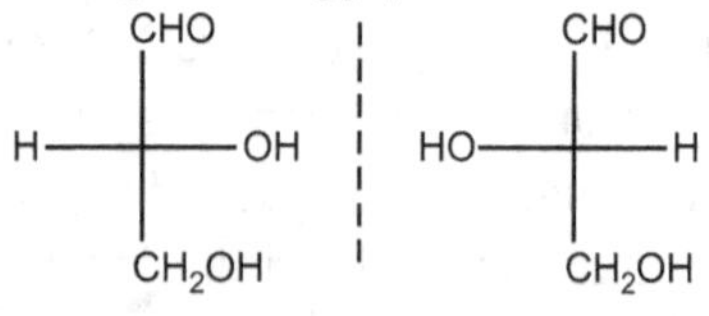

$$
\begin{array}{c}
\text{CHO} \\
\mathrm{H} \!-\!\!-\!\!-\!\!-\! \text{OH} \\
\text{CH}_2\text{OH} \\
\text{R-Isomer}
\end{array}
\qquad
\begin{array}{c}
\text{CHO} \\
\text{HO} \!-\!\!-\!\!-\!\!-\! \mathrm{H} \\
\text{CH}_2\text{OH} \\
\text{S-Isomer}
\end{array}
$$

(ii) Conformational isomerism: The stereoisomers that interconvert through rotation about sigma bonds, is a low-energy process.

Example:

Eclipsed conformation   Staggered conformation

Thus, various types of isomerism could be summarized and detailed as follows:

Isomerism

Structural isomerism   Stereoisomerism

Chain Positional Functional Tautomerism   Configurational   Conformational

5.4 Representation of Conformations of Ethane by (Wedge Formula, Newman, Saw-horse and Fischer Representations)

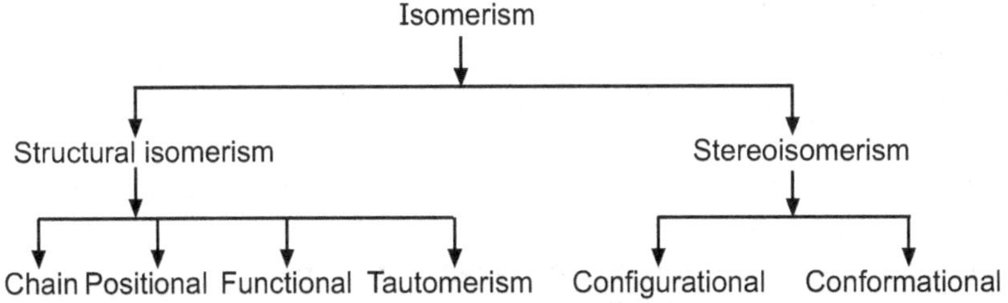

Flying wedge   Sawhorse   Newman

Fischer

Fig. 5.1: Representation of ethane in Flying, Saw-horse, Newman and Fischer's projection formula

The reverse may be carried out to obtain Flying wedge formula from Fischer's formula.

5.5 Conformation of Alkanes

5.5.1 General

We know C–C sigma bond in ethane is axially symmetrical, hence one can easily rotate C-C bond around its axis without changing the bond angles and bond lengths. This rotation changes the relative position of atoms or groups in space attached to these carbons giving different structures. The various shapes or structures that a molecule can adopt by rotation about C-C single bond are called conformation.

The various structures obtained are due to one type of strain called torsional strain which is "repulsive interaction between bonds (bonding electrons) on adjacent atoms. This leads to restricted rotation, such interaction is known as torsional strain." The angle of rotation about the single bond is called the dihedral angle (θ) sometimes called torsional angle. This is an important stereo chemical parameter used to study the conformational analysis. The potential energy relationship between conformations can be easily discussed in the forms of dihedral angle which is the angle generated by two intersecting planes e.g. in the Newman projection of the eclipsed conformation of ethane the angle between (dihedral) two H-C-C planes is 0 while in a staggered conformation it is $60°$ or $180°$. The angle of rotation about the single bond is called the dihedral angle.

At room temperature the rotation about single bond is much faster that lead to millions of different structures which are separated by definite energy barriers. Thus *"the stereoisomers separated relatively by low energy barrier (<60 kJ/mol) that are easily interconvertible at room temperature without bond braking are called conformational isomers or conformers or rotational isomers or rotamers"*.

5.5.2 Conformational Analysis

A study of various conformations of a compound and their relative stabilities is called conformational analysis.

Theoretically, infinite number of conformations is possible even for a simple molecule. If the rotation about the single bond is free then the molecule can assume all these conformations with complete freedom. Pfizer, however, found that there is some energy barrier for such rotation around the single bond and it is not absolutely free. This leads to different conformations with different energies and are called conformers.

5.5.3 Conformational Analysis of Ethane

The various conformation of ethane is best represented by Newman representation.

Types of conformations: In ethane molecule, two methyl groups are attached by a single bond. Due to rotation about the single bond, it can have has two extreme conformations called the "eclipsed" and "staggered" conformations which are obtained by rotating C_1-C_2 bond by 60° intervals.

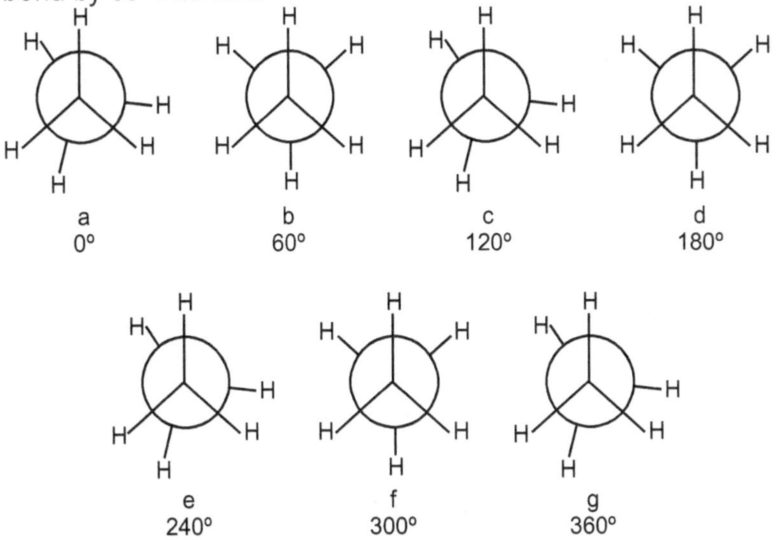

<table>
<tr><td>a
0°</td><td>b
60°</td><td>c
120°</td><td>d
180°</td></tr>
<tr><td>e
240°</td><td>f
300°</td><td>g
360°</td><td></td></tr>
</table>

Conformations of ethane

(a) Eclipsed conformation: In these conformations the atom or groups attached to near/front carbon (C_1) completely block the view of the far/back carbon (C_2). In ethane there are three identical eclipsed conformations (a, c and e) occur at three different dihedral angles viz. 0°, 120° and 240°. In eclipsed conformation, the adjacent C – H bonds are exactly opposite to each other. Thus, when viewed along the axis of the carbon atoms, the three H-atoms of the rear carbon atom will be exactly behind those of the front carbon atom. In this conformation hydrogens are closest with each other, hence the highest energy conformation.

(b) Staggered conformation: Those conformations in which the atoms or groups attached to far/back (C_2) carbon appear in the gaps between the near/front carbon (C_1). In ethane there are three identical staggered conformations (b, d and f) occur at three different dihedral angles viz. 60°, 180° and 300°.

The potential energy of these conformations are less than eclipsed and thus more stable (about 3 kcal/mol less). In this conformation, hydrogens are farthest from each other, hence the lowest energy conformation.

(c) Skew conformations: All other conformations are referred as skew conformations.

Relative stabilities of conformations (The potential energy diagram):

Fig. 5.2 shows the potential energies of different conformations of ethane.

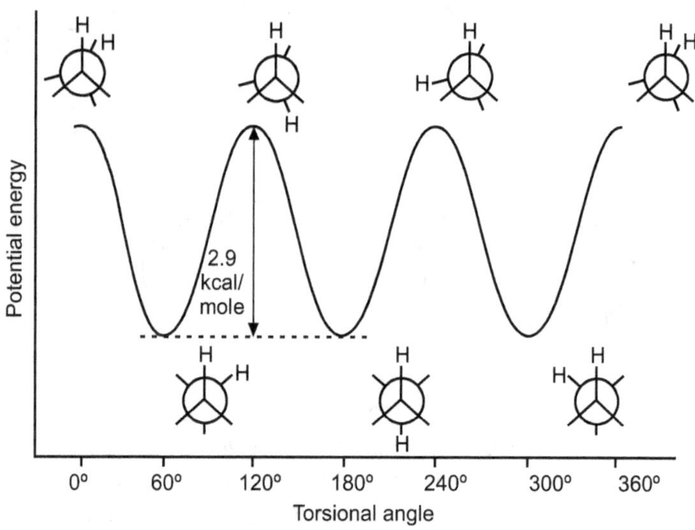

Fig. 5.2: Potential energy diagram for the conformations of ethane molecule

A profile changes with energy maxima (peaks) and minima (valleys) periodically. The diagram shows that the staggered conformation has low energy, while eclipsed has highest energy. This means that the staggered conformation is more stable than eclipsed one by energy 3 kcal/mol or 12 kJ/mol i.e. the energy difference between the two conformers called barriers of interconversion is about 3 kcal/mol. This energy required for rotation about the C-C bond is small and at room temperature the thermal energy from the surrounding (20 kcal/mol) is sufficient to cause rapid interconversion of two conformations. At one given time interval, majority of the ethane molecules are found in staggered and least number in eclipsed conformations.

5.5.4 Conformational Analysis of n-Butane

When we consider n-butane, C_4H_{10}, in which the four carbon atoms are in straight chain, and focus our attention on the middle C – C bond, we see molecule similar to ethane, but with a methyl group replacing one hydrogen on each carbon. As in ethane, the eclipsed conformations have higher torsional energies and hence less stable than the staggered conformations.

But in n-butane due to the presence of methyl groups, there are several different staggered conformations due to torsional strain as well as the steric strain.

Types of conformations: In butane the rotation of central C-C bond (C_2 – C_3) is sterically hindered due to repulsive interactions of large $–CH_3$ groups. Thus while rotating the C_2 – C_3 bond by 60° intervals, it gives following (Fig. 5.3) different types of conformations.

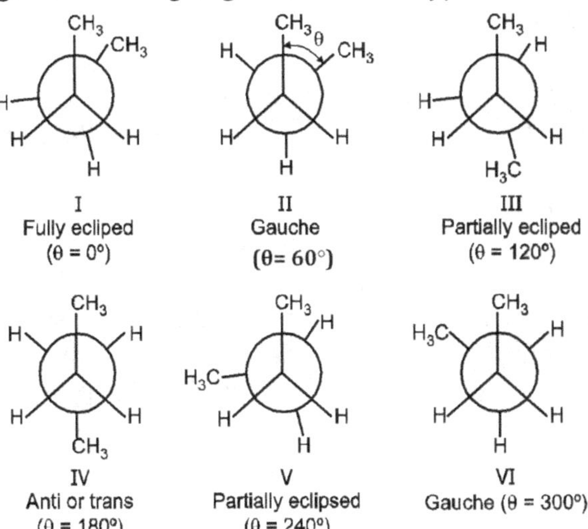

Fig. 5.3: Conformations of n-butane

(A) Eclipsed conformations:

The three conformations (I), (III) and (V) at dihedral angles $0°$, $120°$ and $240°$ are eclipsed conformation. These are again sub-classified into two types:

(i) **Fully eclipsed/syn-periplanar:** This is represented as I conformation at dihedral angle $0°$ in which the rear $-CH_3$ group is eclipsed by front $-CH_3$ group and the two rear H atoms by two front H atoms.

(ii) **Partly eclipsed/anticlinal:** The remaining two eclipsed conformations (III and V) at dihedral angles of $120°$ and $240°$ are identical as they have same P.E. In these conformations the $-CH_3$ group is eclipsed by H atom, hence called partially eclipsed. Actually these conformations are mirror images of each other.

(B) Staggered conformation:

The three other conformations II, IV and VI at dihedral angles of $60°$, $180°$ and $300°$ are staggered conformation and are again sub-classified into two types:

(i) **Gauche/synclined:** The two conformations II and IV are identical as they have same P.E. In these the rear and front $-CH_3$ groups are at angle of $60°$ to each other. Actually these conformations are mirror images of each other.

(ii) **Anti/anti-periplaner:** This is represented as conformation IV at dihedral angle of $180°$. In this the rear and front $-CH_3$ groups are far apart (opposite to each other).

Relative Stabilities of Conformations (The Potential Energy Diagram):

Fig. 5.4 represents the energies of different conformation of n-butane. The diagram shows that;

(a) In fully eclipsed conformations, the two $-CH_3$ groups are at closest distance so they produce more steric strain and cause steric repulsion. Further the close proximity of many C-H bonds leads to electronic repulsion to produce torsional strain. The combined effect of these two renders this conformation highly unstable with P.E. about 4.5-6.1 kcal/mol or 19 kJ/mol, hence this is least favored and least stable conformation among all others.

(b) In partially eclipsed conformation, the bulky $-CH_3$ group is eclipsed with H-atom due to this the steric and torsional strain are comparatively lesser than fully eclipsed. So it is relatively more stable containing less energy by about 3.8 kCl/mol (16 kJ/mol).

(c) In gauche conformation, two $-CH_3$ groups are at $60°$ dihedral angle, hence the steric strain is less as compared to fully eclipsed form, hence they are more stable with P.E. about 0.9 kcal/mol (4 kJ/mol).

(d) In anti-conformation two bulky $-CH_3$ groups are far apart and there is no steric strain. As a result this conformation is most stable of all conformations of n-butane.

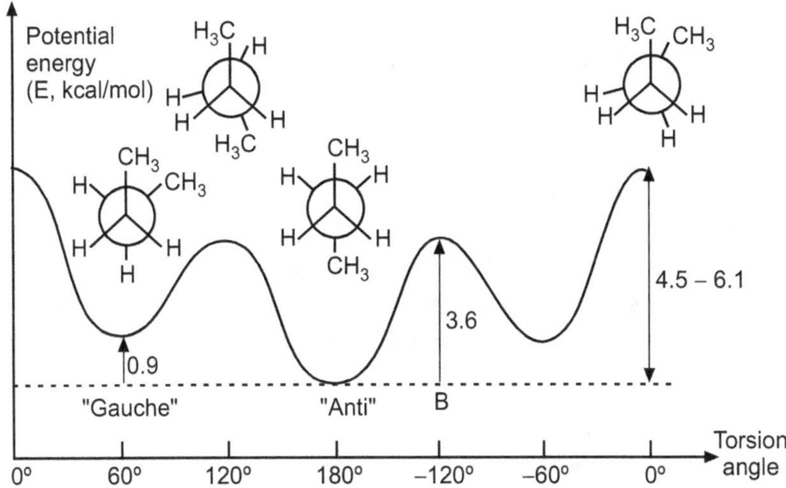

Fig. 5.4: Potential energy diagram for the conformations of n-butane

5.5 Stability and Conformations of Cyclohexane

5.5.1 Stability of Cyclohexane

Cycloalkanes are cyclic hydrocarbons, which form a homologous series with general formula C_nH_{2n}, whose first member is cyclopropane (n=3), next cyclobutane (n=4) and so on. Cyclohexane (n=6) has infinite number of conformations. In order to understand how, let us first understand why the stability of the cyclic rings is as follows:

Cyclopropane < Cyclobutane < Cyclopentane < Cyclohexane

Let us first consider cyclopropane, cyclobutane and cyclopentane as shown in Fig. 5.5.

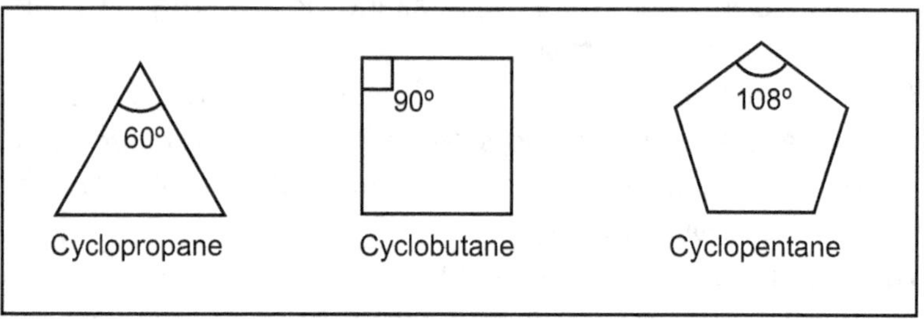

Fig. 5.5: Members of cycloalkanes family

Stability order of above three cycloalkanes can be explained in terms of following theories.

1. Baeyer's Strain Theory:

It states that *the higher the angular/Baeyer's strain, the lesser is the stability of the cyclic ring.* Angular strain can be understood as follows: Although all the carbon atoms of these cyclic rings are sp^3 hybridized and thus normally the angles tend to be around 109°28'. But the cyclopropane and cyclobutane have bond angles of 60° and 90° respectively. This deviation results in angular or torsional strain.

The bond angles of the tetrahedral (sp^3) carbon are about 109°28'. Any phenomenon that tends to increase or decrease this value adds strain to the molecule and increases its energy.

Magnitude of angular strain is given as,

$$\text{Angular strain} = \left| \frac{1}{2} (109° \ 28' - x) \right|$$

where, x = bond angle in the cyclic rings like 60, 90, 108 etc.

The greater the deviation from the normal tetrahedral angle, the greater will be the strain on the molecule. This leads to lesser stability. Baeyer calculated angle of distortion and total strain for each ring (Table 5.1).

Table 5.1: Angle strain and total strain in cycloalkane rings

Compound	N	Angle strain at each CH_2	Heat of combustion (kcal/mol)	Total strain (kcal/mol)
Cyclopropane	3	49.5	499.9	27.7
Cyclobutane	4	19.5	655.9	26.3
Cyclopentane	5	1.5	793.4	6.5
Cyclohexane	6	10.5	944.8	0.4
Cycloheptane	7	19.1	1108.1	6.3
Cyclooctane	8	25.5	1268.9	9.7
Cyclononane	9	30.5	1429.5	12.9
Cyclodecane	10	34.5	1586.1	12.1
Cyclopenta-decane	15	46.5	2362.5	1.5
Open chain alkane				-

According to Baeyer

(i)   Cyclopropane ($x = 60°$) is the most strained among all cycloalkanes and hence least stable.

(ii)   Cyclobutane has a bond angle of $90°$ and hence more stable as compared to cyclopropane.

(iii)   According to Baeyer, cyclopentane should be the most stable of all the cycloalkanes because the ring angles of a planar pentagon is $108°$ and is are closer to the tetrahedral angle than those of any other cycloalkane.

(iv)   Cyclohexane has a bond angle of $120°$ and hence less stable as compared to cyclopropane.

(v)   Out of these cyclopentane has least angular strain and should be most stable.

(vi)   A prediction of the Baeyer strain theory is that the cycloalkanes beyond cyclopentane should become increasingly strained and correspondingly less stable.

According to above calculations, the order of stability should be:

Cyclopropane < Cyclobutane < Cyclopentane > Cyclohexane

Limitations of Baeyer's Strain Theory:

Some of the inconsistencies in the Baeyer strain theory will become evident as we use heats of combustion (Table 5.1) to prove the relative energies of cycloalkanes. The most important column in the table is the heat of combustion per methylene (CH_2) group. The Baeyer's strain theory could explain the stability and reactivity of smaller rings upto cyclopentane but could not be extended to cyclohexane and higher members of cycloalkanes.

(i)   Baeyer assumed that all cycloalkanes are planar, but most of them adopt a puckered conformation. Cyclic molecules can assume non-planar conformations to minimize angle strain and torsional strain by ring puckering.

(ii) Furthermore, he did not consider the contribution of torsional strain to the overall strain energy of a molecule.

(iii) As observed by heat of combustion and total strain.

    (a) cyclohexane is more stable than other rings.

    (b) cyclopentane and cycloheptane are almost equally stable.

In fact, the same is observed experimentally. But this theory cannot be extended to cyclohexane. Baeyer concluded that cyclopropane would be the most strained followed by cyclobutane. Cyclopentane would be strain free while cyclohexane would show a substantial amount of strain energy. Rings larger than cyclohexane would be highly strained and are not capable of existing. Assuming cyclohexane to be a regular hexagon, the bond angle in it would be $120°$ and there would be angular strain, greater than that in cyclopentane. Thus, the cyclopentane should be more stable than cyclohexane! But in fact cyclohexane is more stable than cyclopentane.

2. Sachse-Mohr Theory (Theory of Strainless rings):

The higher stability of cyclohexane than cyclopentane was explained by **Herman Sachse**. According to him, the cyclohexane try to become non-planar, i.e., acquire puckered structure and thus relieve their angular strain. He also suggested that cyclohexane exists in the strain-free chair and boat forms (Fig. 5.6). This suggestion was not accepted at the time as he could not isolate the two forms. In 1918, Mohr proposed that these two isomers are rapidly interconvertible (rapid equilibration), hence could not be separated.

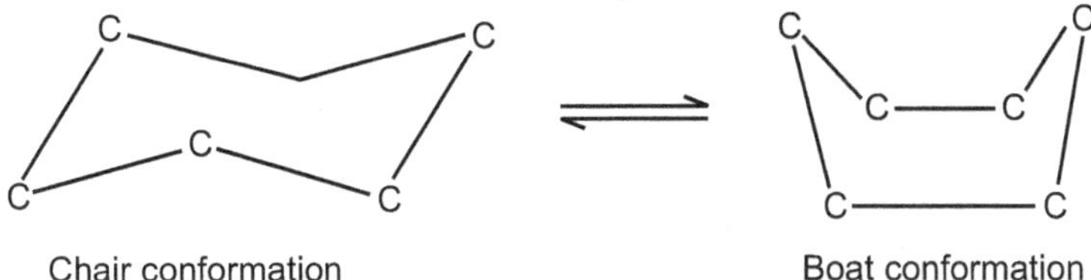

Chair conformation   Boat conformation

Fig. 5.6: Strainless conformations of cyclohexane

The new theory is known as Sachse-Mohr theory or theory of strainless rings. It states that the carbon atoms of cyclohexane and higher cycloalkanes restore normal tetrahedral angle by staying it in different planes and the ring attains a puckered structure. This puckering decreases the P.E of molecule and increases the stability, such a ring system free from angle strain are called strainless rings.

5.5.2 Conformational Analysis of Cyclohexane

There are two important conformations of cyclohexane called chair and boat conformations. They are best represented by Sawhorse. They are freely interconvertible and in between these two there are infinite numbers of gauche or skew conformations.

1. Chair conformation (rigid): This conformation looks like chair and contains two types of H.

Types of Hydrogens: On careful examination of a chair conformation of cyclohexane, we find that the twelve hydrogens are not structurally equivalent. There are two different types of C-H bonds, and thus two different types of hydrogen atoms, axial and equatorial hydrogen atoms.

(a) Axial C-H bonds or H atoms (H$_a$): Out of twelve, six C-H bonds are oriented above and below the plane of the ring (three in each location), and are termed axial because they are aligned parallel to the symmetry axis of the ring. In the figure above, the axial hydrogens are represented as H_a. There are six of these, three upward and three downward bonds, and they alternate up/down around the ring.

(b) Equatorial C-H bonds or H atoms (H_e): The other six of them are located about the periphery of the carbon ring, and are termed as equatorial. These hydrogens are represented as H_e. There are six of them, three of which are "slant up" and three of which are "slant down", again alternating around the ring.

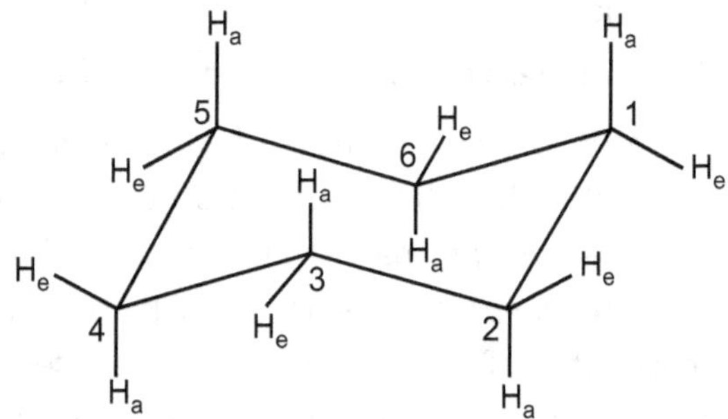

Fig. 5.7: Chair conformation of cyclohexane with different H atoms

2. Boat (Flexible) conformation:

This conformation looks like boat shape and contains four types of H.

Types of Hydrogens: On careful examination of a boat conformation of cyclohexane, we find that the twelve hydrogens are not structurally equivalent. There are four different types of C-H bonds, and thus four different types of hydrogen atoms, flagpole, bow-sprit, quasi-axial and quasi-equatorial hydrogen atoms.

(a) Flagpole C-H bonds or H atoms (H_{fp}): The two C-H bonds on C_1 and C_4 are coming towards each other, and are termed as **flagpole** C-H bonds. These hydrogens are represented as H_{fp}.

(b) Bow-sprit C-H bonds or H atoms (H_{bs}): The other two C-H bonds on C_1 and C_4 are orienting apart to each other and are termed as bow-sprit C-H bonds. These hydrogens are represented as H_{bs}.

(c) Quasi-axial C-H bonds or H atoms (H_{qa}): The four C-H bonds on C_2, C_3, C_5 and C_6 are forms other sides of boat are termed as quasi-axial C-H bonds and are almost parallel to 4the axis. These hydrogens are represented as H_{qa}.

(b) Quasi-equatorial C-H bonds or H atoms (H_{qe}): The other four C-H bonds on C_2, C_3, C_5 and C_6 are forms other sides of boat are termed as quasi-equatorial C-H bonds. These hydrogens are represented as H_{qe}.

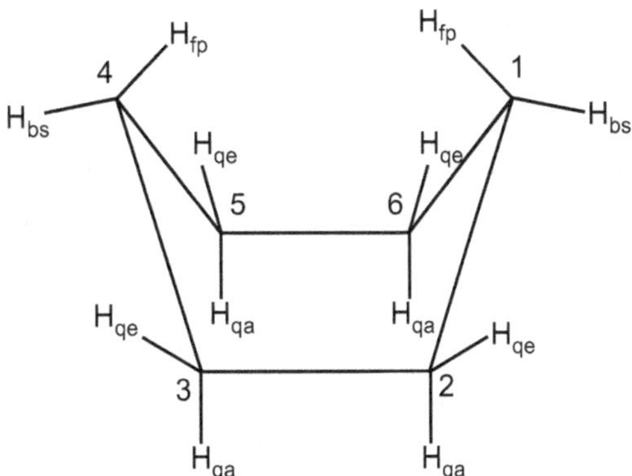

Fig. 5.8: Boat conformation of cyclohexane with different H atoms

Its Newman's representation is shown in Fig. 5.9. The chair conformation has a low torsional strain as seen in a Newman projection.

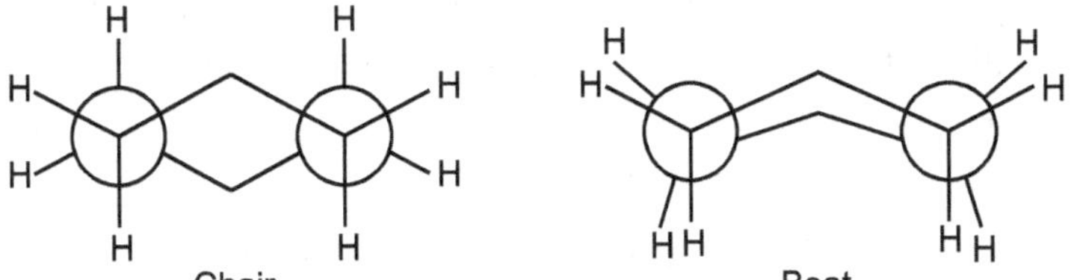

Chair

Boat

Fig. 5.9: Newman's representation of chair and boat conformation of cyclohexane

3. Twist Conformation:

It is more stable than the boat conformation, but less stable than the chair conformation (Refer Fig. 5.10). The flagpole interactions and tensional strain in the boat conformation are reduced in the twist conformer.

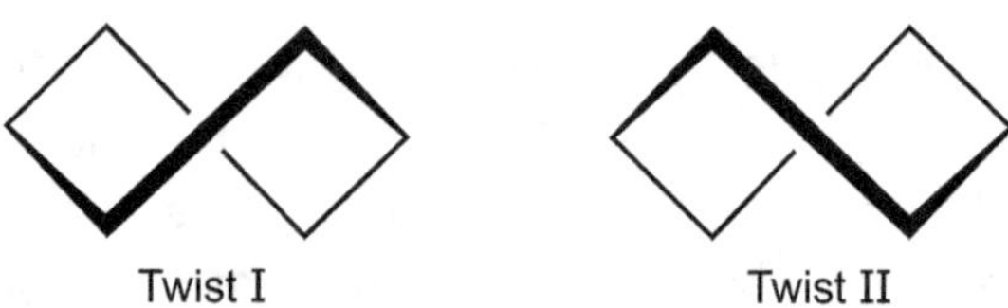

Fig. 5.10: Twist conformations of cyclohexane

Ring flipping or Chair-flipping: The conversion from one chair shape to the other is called ring flipping or chair-flipping (Refer Fig. 5.11). The C-H bonds that are axial in one configuration become equatorial in the other, and vice versa; but their relative positions remain the same. In cyclohexane, the two chair conformations have the same energy, and at 25°C, 99.99% of all molecules in a cyclohexane solution will be in a chair conformation. Since there are two equivalent chair conformations of cyclohexane in rapid equilibrium, all twelve hydrogens have 50% equatorial and 50% axial character.

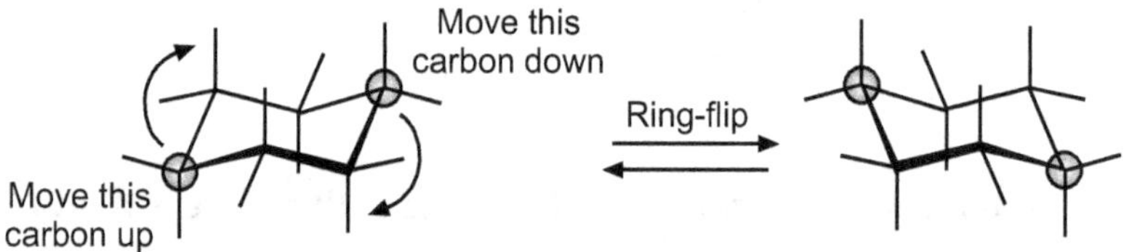

Fig. 5.11: Ring flipping or chair-flipping

Relative stabilities of conformations (The potential energy diagram): The cyclohexane continuously flips from one chair conformation to the other. Approximately 1 million such inter-conversions occur every second. During ring flipping it passes through following different conformations.

Chair 1 → Half Chair 1 → Twist Boat 1 → Boat → Twist Boat 2 →

Half Chair 2 → Chair 2

The change in potential energy as a function of conformation is shown in Fig. 5.12. It is observed that chair conformation contains less strain and hence most stable. While, half chair is highly unstable with energy 46.02 kJ/mol. The boat structure still has two eclipsed bonds and severe steric crowding of two hydrogen atoms on the "bow" and "stern" of the boat. The steric interaction of these two H atoms, known as the **flagpole interaction** also destabilizes the boat conformation. This steric crowding is often called steric hindrance. Torsional strain and flagpole interactions cause boat conformation to have considerably higher energy than chair conformation. Hence the chair form is more stable than the boat form by 44 kJ/mol^{-1}. By twisting the boat conformation, the steric hindrance can be partially relieved.

From figure the relative stabilities of cyclohexane conformations decrease in the following order:

Chair >Twist Boat > Boat > Half Chair

At room temperature, the average energy of cyclohexane is so sufficient that, it overcomes the energy barrier between two chair conformations and so it exists in dynamic equilibrium. Hence it is not possible to separate these conformations of cyclohexane.

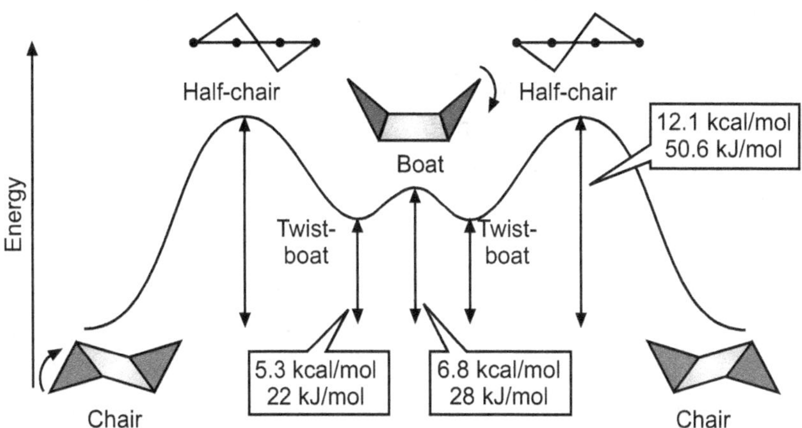

Fig. 5.12: Potential energy diagram for the conformations of cyclohexane

5.6 Conformational Analysis of Monosubstituted Cyclohexanes

The most stable form of a monosubstituted cyclohexane is, like cyclohexane itself, a chair conformation. When a cyclohexane molecule flips between two chair forms, the energy level for both forms is the same. However, if there is a substituent on the ring, the axial and equatorial positions have different spatial arrangements and the two conformations are no longer equivalent.

There are two chair conformations:

1. Axial conformation in which substituent is at axial position.
2. Equatorial conformation in which substituent is at equatorial position.

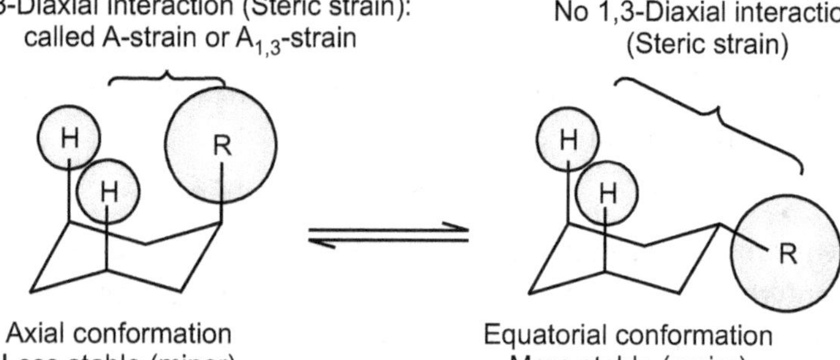

1. Conformational analysis of Methyl cyclohexane:

If hydrogen of cyclohexane is replaced with a methyl group to form methylcyclohexane, the two chair forms are different.

(i) **Equatorial conformation**: In this the methyl group is oriented at equatorial position.

(ii) **Axial conformation:** In this the methyl group is oriented at axial position.

The two chair forms are not present in equal amounts. Experimental evidence shows that at room temperature approximately 95 % of the methyl groups are equatorial, due to the differences in stability of the two conformations. In any equilibrium process, the chair form of methyl cyclohexane having the equatorial methyl group is the more stable. As you see, an axial methyl group is quite close to the axial hydrogens on C_3 and C_5. This distance is less than the sum of the van der Waals radii for two hydrogens (240 pm), so the axial conformation of methyl cyclohexane is destabilized by the strain of the van der Waals repulsive forces (i.e. 1 : 3-diaxial interaction also called $A_{1,3}$-strain).

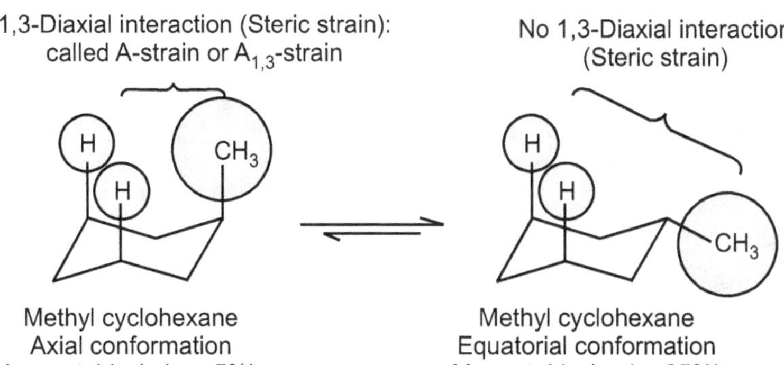

Methyl cyclohexane
Axial conformation
Less stable (minor-5%)

Methyl cyclohexane
Equatorial conformation
More stable (major-95%)

The energy barrier between axial and equatorial conformation is 1.7 kcal/mol, so at room temperature they can interconvert readily (Fig. 5.13). Hence, they cannot be separated from one another.

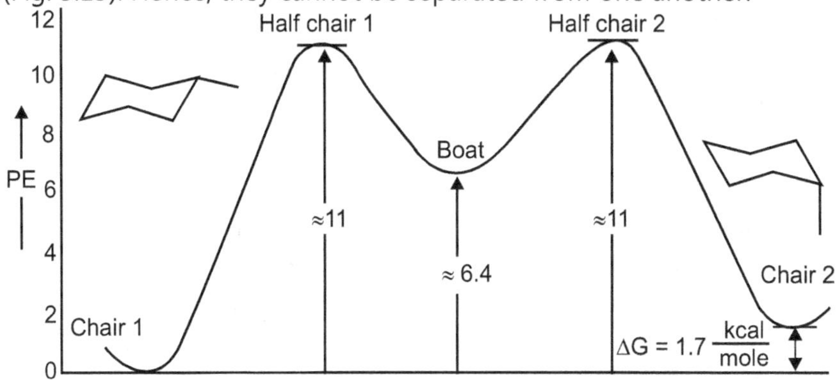

Fig. 5.13: Potential energy diagram for the conformations of methyl cyclohexane

2. Conformational Analysis of Cyclohexanol:

It exists in two conformational isomers.

(i) Axial conformation: In this OH group is oriented at axial position. As you see, an axial OH group is quite close to the axial hydrogens on C_3 and C_5. In this conformation, axial OH group experiences lot of strain due to 1:3-diaxial interaction.

(ii) Equatorial conformation: In this the methyl group is oriented at equatorial position. In this conformation, OH group is oriented at equatorial position, hence there is no 1 : 3 diaxial interaction. Hence equatorial conformation is more stable as compared to axial conformation by 0.87 kcal/mol energy.

Hence the most stable equatorial conformation is present in greater quantities (82 %). The energy barrier between axial and equatorial conformation is 0.87 kcal/mol (Energy profile diagram is similar to Fig. 5.13).

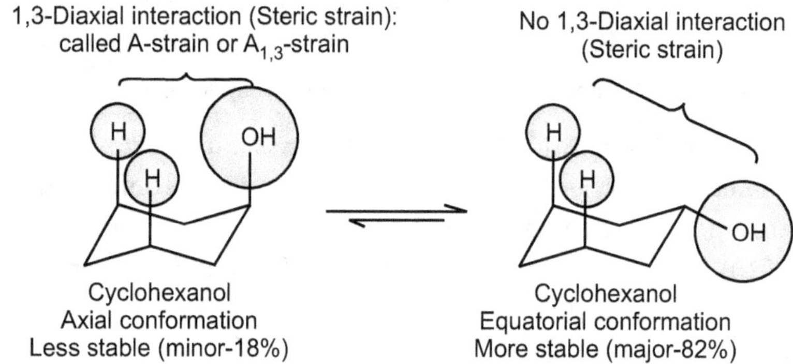

Cyclohexanol
Axial conformation
Less stable (minor-18%)

Cyclohexanol
Equatorial conformation
More stable (major-82%)

Note: Given that oxygen has a larger atomic number than carbon, it is not unreasonable to think that the OH group might be "bulkier" than carbon. When you think about the source of strain in CH_3, however, you realize that it is not necessarily the size of the carbon atom itself but the hydrogens of CH_3 interacting with the axial hydrogens on the ring that lead to strain. Oxygen, having only one hydrogen, can always rotate such that the H is pointing away from the cyclohexane, thereby leading to very little in the way of diaxial interactions with the ring.

3. **Conformational analysis of Bromocyclohexane:**

It exists in two conformational isomers.

(i) Axial conformation: In this Br group is oriented at axial position. As you see, an axial Br group is quite close to the axial hydrogens on C_3 and C_5. In this conformation, axial Br experiences lot of strain due to 1:3-diaxial interaction.

(ii) Equatorial conformation: In this the methyl group is oriented at equatorial position. In this conformation, Br group is oriented at equatorial position, hence there is no 1 : 3-diaxial interaction. Hence equatorial conformation is more stable as compared to axial conformation by 0.43 kcal/mol energy.

Hence the most stable equatorial conformation is present in greater quantities (67%). The energy barrier between axial and equatorial conformation is 0.43 kcal/mol (Energy profile diagram is similar to Fig. 5.14).

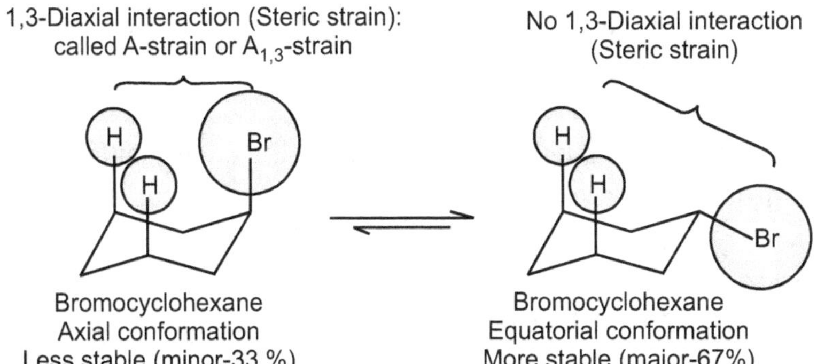

Bromocyclohexane
Axial conformation
Less stable (minor-33 %)

Bromocyclohexane
Equatorial conformation
More stable (major-67%)

Note: Along similar lines one could be forgiven for thinking that Br, being such a heavy and large atom, might exert a large destabilizing influence when in the axial position. However, the difference is only 0.43 kcal/mol, less than that for OH. Why might this be? The answer here is bond length. The average C–Br bond is about 193 pm in length as compared to 150 pm for the C–CH$_3$ in methylcyclohexane. The Br, being farther away, will thus have less interaction with the axial hydrogens.

5.7 Locking of Conformation in *t*-butyl Cyclohexane

Due to the difference in energy between placing a substituent in the axial versus equatorial position, the two chair conformations are no longer equal in energy. As substituent becomes larger, steric interactions with 1, 3 hydrogens increase (due to 1 : 3-diaxial interaction). This interaction can be represented in terms of A-values.

"**A-values**" are a numerical way of rating the bulkiness of substituents on a cyclohexane ring. The "A-value" represents the difference in energy (kcal/mol) between the equatorial and axial conformation of substituted cyclohexane. The greater the "A-value", the higher the energetic preference for the equatorial position, and the more "bulky" the group is considered. For a monosubstituted cyclohexane, the energy difference between axial and equatorial conformers with a given substituent is known as its A-value.

In the previous point, we saw that adding a methyl group to cyclohexane results in two chair conformers that are unequal in energy. We saw that the equatorial conformation, where the methyl group was equatorial is the most stable i.e. 1-methylcyclohexane exists as a 95:5 ratio at room temperature.

The next logical question is this. What is the energy difference for other groups? For example, what happens when we substitute ethyl, *i*-propyl, *t*-butyl ? How is the equilibrium affected?

As groups get larger, the A-value increases.

Table 5.2: A-values for monosubstituted cyclohexanes

R	$A_{1,3}$ kcal/mol	Ratio (equatorial : axial)
CH_3	1.74	95 : 5
CH_2CH_3	1.75	96 : 4
$CH(CH_3)$	2.15	97 : 3
$C(CH_3)_3$	4.90	99.97 : 003 **nearly 100 % *t*-butyl cyclohexane molecules are present in equatorial form**

The *t*-Bu group is too bulky. In axial conformation, a 1,3-diaxial interaction between one of the methyl group and an axial C-H is unavoidable. Hence, axial t-butyl groups are strongly disfavoured. Thus, t-Bu has tendency to remain in equatorial position and resists it to go at axial position. Hence, t-butyl group is called as "*locking group*" as it locks the conformation at equatorial form. This phenomenon is called '*locking of conformation.*'

1,3-Diaxial interaction (4.90 kcal/mol)   No 1,3-Diaxial interaction

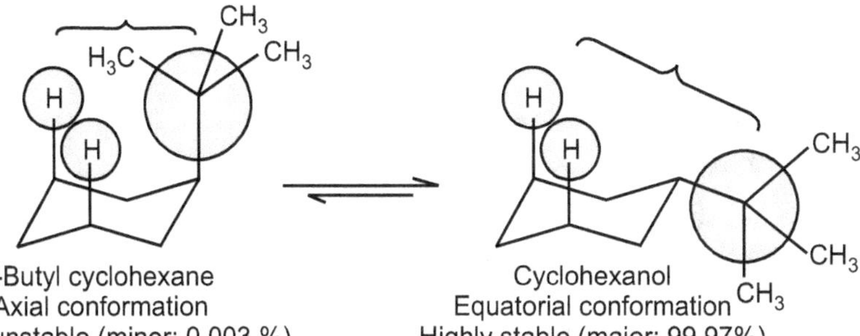

t-Butyl cyclohexane   Cyclohexanol
Axial conformation   Equatorial conformation
Highly unstable (minor: 0.003 %)   Highly stable (major: 99.97%)

Because of this huge difference in stability, we often consider a *t*-butyl substituent on cyclohexane to be '*locked*' in the equatorial position.

Exercises

(A) Define the following terms with example:

1. Conformational isomers.
2. Anti-conformation of butane.
3. Dihedral angle.
4. Rotamers.
5. Strainless ring.
6. Flag-pole hydrogens.
7. Ring flipping
8. Gauche conformation
9. Baeyer's strain

(B) Short Answer Questions:

1. Explain the following terms with suitable examples:

   (a) Baeyer's strain

   (b) Chair conformation of cyclohexane

   (c) Ring flipping

   (d) Less stability of boat conformation of cyclohexane

   (e) Theory of strainless rings

   (f) Locking of conformations

2. Draw different conformations of ethane and explain their stabilities.

3. Draw different conformations of n-butane.

4. Why is staggered conformation more stable than eclipsed conformation?

5. Write notes on :

    (a) Ring flipping   (b) Sachse-Mohr theory

    (c) Baeyer's strain theory

(C) Long Answer Questions:

1. Explain Baeyer's strain theory. What are its limitations?

2. Draw different conformations of cyclohexane and explain its stabilities.

3. Draw different conformations of n-butane and explain their stabilities.

4. Why chair conformation is more stable than boat in cyclohexane? Comment on this using potential energy diagram.

5. What is different projection formulae used to represent stereoisomes of ethane?

6. What is meant by conformations? Explain different conformations of cyclohexane.

7. Assign name to the following conformation :

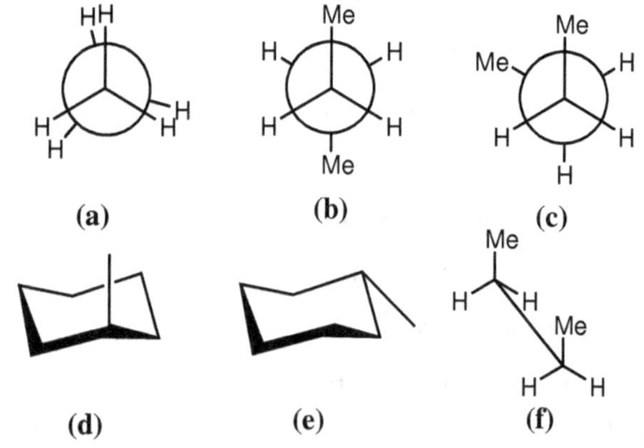

8. Compare the stability of two conformations with reason :

    (i) Eclipsed and staggered conformation of ethane.

    (ii) Eclipsed and staggered conformation of butane.

    (iii) Chair and boat conformation of cyclohexene.

9. What is conformational isomerism? Discuss conformational isomerism in butane.

(D) Choose the Correct Alternative for each of the following and rewrite the sentence:

1. The compound butane shows ...... isomerism.

    (a) geometric   (b) conformational

    (c) optical   (d) none of these

2. The Baeyer's angle strain for following ring system is ......

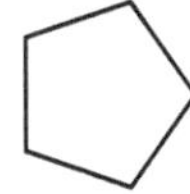

    (a) 49.5   (b) 19.5

    (c) 1.5   (d) 10.5

3. Which of the following is the correct order of stability of ring system?

    (a) 5<4<3   (b) 6<5<7

    (c) 4<5<6   (d) 3<5<4

4. The correct representation of anti-conformation of n-butane is ......

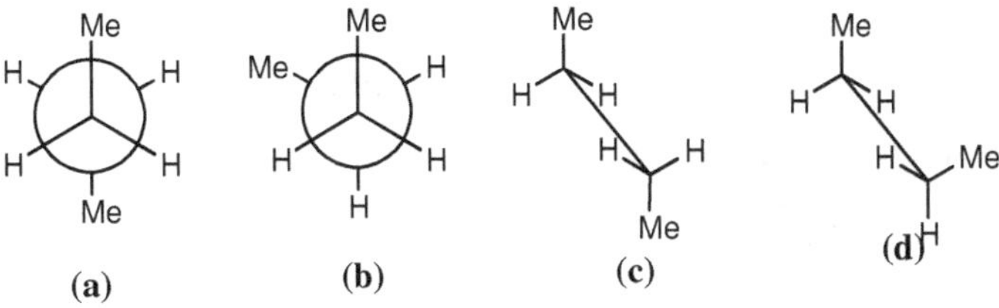

    (a) (a) only   (b) (c) only

    (c) (b) and (d)   (d) (b) and (c)

5. How many number of quasi-axial hydrogens does the boat conformation of cyclohexane have?

(a) 2   (b) 3

(c) 4   (d) 5

6. Which of the following statements best describe the stability of cyclohexane?

(1) axial chair conformation is less stable due to the presence of 1,3-diaxial interactions.

(2) boat conformation is less as compared to chair due flagepole-flagepole interaction.

(3) half chair is most stable conformation of cyclohexane.

(4) half chair is least stable conformation of cyclohexane.

(a) all

(b) 1, 2 and 3

(c) 1, 2 and 4

(d) 1 and 4

7. Choose the false statement about the following :

(1) eclipsed conformation is less stable due to the presence of bonding interactions.

(2) in gauche conformation of n-butane the dihedral angle is 60°.

(3) eclipsed conformation of n-butane is less stable as compared to eclipsed conformation of ethane.

(4) half chair is most stable conformation of cyclohexane

(a) all

(b) 1, 2 and 3

(c) 4

(d) 1 and 4

7. The structures below are:

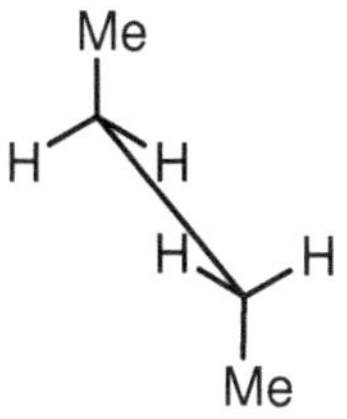

(c) 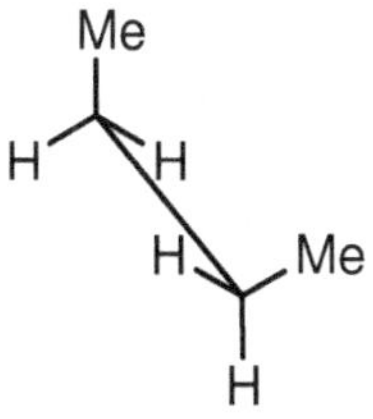   **(d)**

    (a) not isomers   (b) conformational isomers.

    (c) cis-trans isomers   (d) both (b) and (d)

8. Among the butane conformers, which occur at energy minima on a graph of potential energy versus dihedral angle?

(a) gauche only

(b) fully eclipsed

(c) anti

(d) partly eclipsed only

9. Which of the following correctly ranks the cycloalkanes in the order of increasing ring strain per methylene (CH_2 group) ?

(a) cyclopropane < cyclobutane < cyclohexane < cyclopentane

(b) cyclopentane < cyclopropane < cyclobutane < cyclohexane

(c) cyclopentane < cyclohexane < cyclobutane < cyclopropane

(d) cyclopropane < cyclopentane < cyclobutane < cyclohexane

10. Which of the following correctly lists the conformations of cyclohexane in the order of increasing energy?

(a) chair < boat < twist < half-chair

(b) half-chair < boat < twist < chair

(c) chair < twist < boat < half-chair

(d) half-chair < twist < boat < chair

11. Which of the following is the correct order of stability of the four distinct conformations of n-butane ?

   (a) staggered > partially eclipsed > gauche > fully eclipsed

   (b) gauche > staggered > partially eclipsed > fully eclipsed

   (c) staggered > gauche > partially eclipsed > fully eclipsed

   (d) fully eclipsed > staggered > partially eclipsed > gauche

12. The isomers which can be interconverted through rotation around a single bond are ......

   (a) enantiomers   (b) diastereomers

   (c) conformers   (d) positional isomers

Answer: All (c) are correct.